Fieldwork as Failure

Living and Knowing in the Field of International Relations

EDITED BY

KATARINA KUŠIĆ AND JAKUB ZÁHORA

E-INTERNATIONAL
RELATIONS
PUBLISHING

E-International Relations
www.E-IR.info
Bristol, England
2020

ISBN 978-1-910814-53-6

Production: William Kakenmaster
Cover Image: Risaravut via Shutterstock

A catalogue record for this book is available from the British Library.

E-IR Edited Collections

Series Editors: Stephen McGlinchey, Marianna Karakoulaki & Agnieszka Pikulicka-Wilczewska
Books Editor: Cameran Clayton
Editorial assistance: Clotilde Asangna, Assad Asil Companioni, Shelly Rai Mahajan, Alina Williams

E-IR's Edited Collections are open access scholarly books presented in a format that preferences brevity and accessibility while retaining academic conventions. Each book is available in print and digital versions, and is published under a Creative Commons license. As E-International Relations is committed to open access in the fullest sense, free electronic versions of all of our books, including this one, are available on our website.

Find out more at: http://www.e-ir.info/publications

About the E-International Relations website

E-International Relations (www.E-IR.info) is the world's leading open access website for students and scholars of international politics, reaching over 3 million unique readers annually. E-IR's daily publications feature expert articles, blogs, reviews and interviews – as well as student learning resources. The website is run by a registered non-profit organisation based in Bristol, UK and staffed with an all-volunteer team of students and scholars.

Abstract

This volume aims to unsettle the silence that surrounds fieldwork failure in both methods training and academic publications. While fieldwork has gradually evolved into standard practice in IR research, the question of possible *failures* in field-based knowledge production remains conspicuously absent from both graduate training and writing in IR. This volume fills that lacuna by engaging with fieldwork as a site of knowledge production and inevitable failure. It develops methodological discussions in IR in two novel ways. First, it engages failure through experience-near and practice-based perspectives, with authors speaking *from* their experiences. And secondly, it delves into the politics of methods in IR and the discipline more generally to probe ways in which the realities of research condition scholarly claims.

Editors

Katarina Kušić is currently an ESRC Postdoctoral Fellow at the Department of International Politics at Aberystwyth University. She is particularly interested in fieldwork-based methods, conversations between studies of South East Europe and postcolonial and decolonial thought, and liberalism as politics of improvement.

Jakub Záhora holds an M.A. degree from SOAS and a Ph.D. in International Relations from Charles University, Prague where he now works as a lecturer. His doctoral thesis investigated the depoliticisation of everyday life in Israeli settlements in the West Bank through material and visual configurations. Záhora's research interests cover the Israeli-Palestinian conflict, political ethnography, and critical approaches to security.

Acknowledgments

This book came out of conversations about the dangers of 'solitary' research – both at home and in fieldwork. We accordingly have many to thank. We would like to thank the European International Studies Association for funding the Early Career Researchers workshop in Prague where the volume started. Once in Prague, Berit Bliesemann de Guevara and Xymena Kurowska not only offered insightful advice and discussion points, but also encouraged and helped the preparation of the manuscript – they provide an excellent example of what caring mentorship can be. And finally, we also want to thank the authors who so readily took up this 'early career' project and allowed us to figure things out as we went.

All fieldwork depends on people ready to gift their time and energy to researchers – the sheer number of those people involved in the 13 chapters presented in this book is dizzying. There is no way to thank them all but considering them here is a good reminder of the collective efforts and mutuality that research entails.

Katarina would like to especially thank the Department of International Politics at the University of Aberystwyth for providing her with workspace during the summer of 2019 while this volume was being prepared.

Jakub would like to thank the Department of International Relations at Charles University and especially to Ondřej Ditrych for allowing him enough freedom to pursue this project alongside other responsibilities.

Contributors

Berit Bliesemann de Guevara is a Reader in International Politics and the Director of the Centre for the International Politics of Knowledge at the Department of International Politics, Aberystwyth University. Her current research explores ways and problems of knowing in international politics. She has failed in fieldwork in Bosnia-Herzegovina, Kosovo, Berlin, New York, Washington, Myanmar and most recently Colombia, but is today grateful for the productive ruptures that these failures have enabled.

Lydia C. Cole is a Postdoctoral Researcher at University of Durham on the project 'The Art of Peace: Interrogating Community Devised Arts-Based Peacebuilding'. She has previously held the position of Associate Lecturer at University of St. Andrews and received her Ph.D. from Aberystwyth University in 2018. Her research spans several fields of inquiry including feminist international relations, peace and conflict studies, visual arts, and transitional justice. She is particularly interested in the uses of ethnographic, artistic, and creative research methodologies.

Jan Daniel is a Researcher at the Institute of International Relations, Prague. In 2017, he obtained his Ph.D. in International Relations from Charles University. He studies the links between narratives of peace and security and their manifestations in local practices of peacekeeping and peacebuilding, as well as politics of security in Central Europe, Lebanon and Syria. He has published on the political sociology of peacekeeping and narratives of hybrid warfare.

Sezer İdil Göğüş is currently a Doctoral Researcher at Peace Research Frankfurt (PRIF) and a Ph.D. student in the Social and Cultural Anthropology Department at Goethe University Frankfurt. Her research focuses on the institutional forms and imaginary concepts used by the Turkish government on one side and people of Turkish origin on the other side to justify and position their political activities in Germany.

Johannes Gunesch is a Ph.D. student at Central European University in Budapest. His ethnographic work seeks to trace the personal and political resonance of the Egyptian uprising, which was sparked and simultaneously circumscribed by capitalist development.

Danielle House is working as a Postdoctoral Researcher on a HERA funded project looking at experiences of death for migrants and minorities in North West Europe, in the Department of Geography and Environmental Sciences, University of Reading. She completed her Ph.D., which explored memory and memorialisation of people disappearing in contemporary Mexico, at the Department of International Politics, Aberystwyth University in 2019.

Xymena Kurowska is an Associate Professor of International Relations at Central European University (CEU). She received her doctorate in political and social sciences from European University Institute in Florence, Italy. She has been a grantee of the European Foreign and Security Policy Studies Programme, CEU principal investigator in Global Norm Evolution and Responsibility to Protect project, Marie Skłodowska-Curie senior research fellow at the Department of International Politics at Aberystwyth University, and serves as academic rapporteur for EU Cyber Direct.

Katarina Kušić is currently an ESRC Postdoctoral Fellow at the Department of International Politics at Aberystwyth University. She is particularly interested in fieldwork-based methods, conversations between studies of South East Europe and postcolonial and decolonial thought, and liberalism as politics of improvement.

Ewa Maczynska is a Ph.D. Candidate at the Doctoral School of Political Science, Public Policy and International Relations at Central European University. Her research focuses on migration and pro-migrant solidarity movements in Europe.

Emma Mc Cluskey is a Research Associate and Teaching Fellow at the Department of War Studies, King's College London. She is the author of *From Righteousness to Far Right: An Anthropological Rethinking of Critical Security Studies* (McGill-Queens University Press; 2019) and is working on the Open Research Area funded project; GUARDINT, Who guards the guardians? She is co-editor in chief of the bi-annual journal Political Anthropological Research in International Social Sciences (PARISS).

Holger Niemann is a Postdoctoral Researcher at the Institute for Peace Research and Security Policy at the University of Hamburg (IFSH). His research currently focuses on the role of norms and practices in the politics of international organisations, most notably the responsibility of the United Nations Security Council.

Amina Nolte is a Research Associate at the Collaborative Research Center at the Justus-Liebig-University of Giessen. She works on the issue of infrastructure and securitisation in Israel/Palestine from a post-foundational perspective.

Desirée Poets is Assistant Professor of Postcolonial Theory at Virginia Tech's Department of Political Science as well as Core Faculty member of the Alliance for Social, Political, Ethical, and Cultural Thought (ASPECT) Ph.D. program.

Renata Summa holds a Ph.D. in International Relations from Pontifical Catholic University of Rio de Janeiro and an M.A. in International Relations from Sciences Po Paris. She teaches at the Institute of International Relations of PUC-Rio. Summa conducted fieldwork in Bosnia and Herzegovina in 2014 and 2015. Her research interests are conflict and post-conflict situations, borders and boundaries, the Balkans and everyday approaches in IR.

Jakub Záhora holds an M.A. from SOAS and a Ph.D. in International Relations from Charles University, Prague where he now works as a lecturer. His doctoral thesis investigated the depoliticisation of everyday life in the Israeli settlements in the West Bank through material and visual configurations. Záhora's research interests cover the Israeli-Palestinian conflict, political ethnography, and critical approaches to security.

Contents

Introduction

Fieldwork, Failure, IR

KATARINA KUŠIĆ AND JAKUB ZÁHORA

This volume aims to unsettle the silence that surrounds fieldwork failure in both methods training and academic publications. Speaking from the practice of Ph.D. research to (postgraduate) researchers embarking on fieldwork-based research projects, it seeks to problematise the notion of fieldwork-based methods as mere instruments for data-gathering. In doing so, it joins volumes that deal with practical aspects of fieldwork (Cerwonka and Malkki 2007; Nordstrom and Robben 1995); more recent reflections on fieldwork in specific contexts (Kapiszewski, MacLean, and Read 2015; Bliesemann de Guevara and Bøås 2020; Glasius et al. 2018); and novel experiments with forms of writing (Pachirat 2018). We further build on works that have successfully brought methodological discussion to the centre of International Relations (IR) debates: feminist reflections on the political and ethical investments of researchers (Wibben 2016; Ackerly, Stern, and True 2006); thinking about the role of methods in critique (Salter and Mutlu 2012; Aradau et al. 2015); probing of the ethical dilemmas of immersion (Dauphinée 2007); and calls for reflexivity (Steele, Gould, and Kessler 2019). The authors assembled here complement this rich literature by addressing the specificities of conducting fieldwork within IR and speaking *from*, rather than about, the difficulties of living and knowing in the field.

Learning Methods

We, the editors, thought ourselves to be 'responsible' IR scholars when we started our Ph.D.s in 2014. We took methods seriously: we read the heated debates on ethnography, we were aware of the colonial underpinnings of the terms 'field' and 'home', and we expected fieldwork to be power-laden and challenging. We approached these problems solemnly, reflectively, and with trepidation. And, importantly, we met at a place that promised to give us the tools to solve them – a summer school on ethnographic methods in Ljubljana.

The idea of method that underpins the possibility of a 'methods school' is one of a *recipe*. To use John Law's (2007, 9) formulation, as students we are told that doing methods 'properly' will ensure 'a healthy research life'. Approached in this way, methods are a series of *do* and *don't* instructions. They are given as advice, summaries of best practices, and lists of things we should do to turn our research design into knowledge. As we shared our fears and excitement with colleagues in Ljubljana, we were relieved by these promises: if we followed the instructions, we too would become successful researchers.

In addition to this 'healthy' research, fieldwork-based methodologies promise *exciting* research. Within IR, they were introduced with worthy aims: to 'resolve the aporias of textual representation', work towards 'emancipation', and cultivate reflexivity (Vrasti 2008, 284). The lure grew stronger as we learned about the political investments of methods themselves: how we approach the field, design our research, position ourselves among our collaborators, and the ways we write are all processes shaped by power. Methods are an opportunity to situate ourselves in relation to that power, they allow us to enact some and disrupt other worlds (Aradau and Huysmans 2014). The excitement of confronting these politics was as attractive as the promise of the many secrets we imagined finding in the field.

After our methods school in 2015, we had not seen each other again until November 2017 when we both attended a workshop in London. In this new meeting, we were not supposed to learn about methods, but present findings that we had reached through practicing them in the two years spent in our respective fields, Katarina in Serbia and Jakub in Israel/Palestine. Catching up, we quickly concluded that we had not succeeded in doing 'real ethnography', and, despite perhaps earning us degrees, our fieldwork had mostly been a 'failure'.

In subsequent informal chats with colleagues and friends, 'failure' increasingly appeared to be part and parcel of fieldwork methods and knowledge production more generally. Stories multiplied: botched interviews, sexual harassment, broken limbs, ruined relationships, political inadequacy, inescapable guilt – failures were everywhere, yet conspicuously absent in public debates about fieldwork, publications, and conferences.

Our informal conversations continued at an Early Career Researchers Workshop at the EISA convention in Prague. The workshop demonstrated the potential of talking about failure: it allows us to be more transparent about the many material and emotional factors that shape our research, exposes long-standing academic conventions that form academic subjectivities, and provides an opportunity to challenge academia's obsession with productivity and the narrative of disembodied research. This volume is an outcome of this open, trustful sharing.

Failing in/to... what?

Thinking about failure in academia is confusing. As Sarah Naumes (2015, 827) notes, 'to admit to failure may as well be a cardinal sin in the academy'. At the same time, failure is emerging as a political concept capable of challenging the neoliberal imperative of success and productivity. Failures are retold in efforts to 'provide some perspective' by publicising the many rejections that make an average academic career.[1] Recently, Laura Sjoberg encouraged us to embrace failure in critical scholarship, as critique 'is always and already failing and failed' (Sjoberg 2018). Even within methodological discussions, failure can mean a variety of different things: from projects that were prevented from ever being (Smith and Delamont 2019), to those that changed dramatically in response to changing conditions in the field (Kurowska 2019; Daigle 2016).

After many conversations, we recognise that the failure we observed in the London workshop connects the affective experiences of the researcher with larger political and epistemological investments of our research. We do fieldwork looking for different perspectives, political relevance, and engaged writing. In her review of IR ethnographies, however, Wanda Vrasti (2008, 284) points out that these aims do not only remain unachieved, but having those expectations in the first place is proof that IR scholars have *failed* to understand what ethnography really is (cf. Lie 2013). Many of the failures recounted in this volume relate to these great expectations, despite all of us having read both Vrasti's work and the anthropological debates that inspired it.[2]

No matter the management of expectations and the humility that we profess, most of us do fieldwork because we believe that the stories we are told by those we meet 'in the real word' are worth telling. Ethnography is animated by a commitment to an epistemology that recognises our necessarily limited and partial positionality, and by extension the political value of attending to others' perspectives (see Haraway 1988; Harding 1992b; 1992a). Even while being aware that fieldwork is in no way free of the problems of representation and the powers that shape it (Dauphinée 2007), we travel to field sites because it

1 The CV that went viral belongs to Johannes Haushofer at Princeton (https://www.princeton.edu/~joha/Johannes_Haushofer_CV_of_Failures.pdf), but the idea was first developed in Melanie I. Stefan's (2010) article in *Nature*.

2 We have in mind the 'crisis of representation' debates that started in social anthropology in 1986 with the publication of *Writing Culture* (Clifford and Marcus 1986) and *Anthropology as Cultural Critique* (Marcus and Fischer 1986).

is an exhilarating encounter with other perspectives that challenge our concepts, change our directions, and make us confront first-hand the power that we study.

Mobilising these perspectives in our research and our knowledge claims necessarily means making people and things legible: we note and transcribe, draft and structure, analyse and write, follow up and reconceptualise in order to make the social world we encounter comprehensible to us and our peers. This aspiration to represent sets us up for a complex failure. It inextricably connects us with a desire for mastery: we represent in order to make the 'authenticity' of being there legible to our epistemic communities, even when we know that any such claim to authenticity is impossible.[3] The stories we tell *necessarily* fail: they are incomplete, situated, and imbued with the power of our own interpretation (Page 2017; Daigle 2016). In short, as we do not have privileged access to ourselves *or* our interlocutors, all our understandings have to be accompanied by an 'awareness of fracture and partiality' (Dauphinée 2016, 48).

While the dismantling of the ideal of mastery and representation is an ongoing and never-finished process, we also fail on a daily basis when dwelling in the field. Rebecca Hanson and Patricia Richards (2019) have recently shown how these everyday experiences are powerfully shaped by three 'ethnographic fixations': solitary research, danger, and intimacy. Even after decades of work that feminist, postcolonial, and critical race theorists have done in academia, these fixations still shape not only what happens to us in the field and the wider academe, but also how we interpret it. Drawing on Hanson and Richards' volume, we argue that both the failures and the absence of these failures in talk and writing say something about the state of our discipline and academia more broadly.

The failures we discuss in this volume are not simple rejections, closures, or endings: they are continuous negotiations in the practice of doing and writing research. In their focus on fieldwork experiences, the contributions are then similar to 'confessionals' – tales in which we are told 'what really happened' in the field and the many failures that accompanied fieldwork (Van Maanen 2011; Rabinow 1977; Jemielniak and Kostera 2010; Schatz 2009; Thomson, Ansoms, and Murison 2013; Yanow and Schwartz-Shea 2006). In addition to 'telling the story', the chapters also uncover and challenge the disciplining structures of science that need to be both navigated and challenged. In this volume, we consider them together to highlight the power that disciplinary expectations have and the ways in which they guide our positioning as

---

3 In her volume *The Queer Art of Failure*, Judith Halberstam (2011, 2) connects this idea of success to the discipline(ing) of social sciences: staying within well-defined parameters helps maintain order and depends on making things legible.

researchers – even when we are explicitly warned against such illusions.

In discussing fieldwork failures, we might be once again setting a goal that is out of our reach: to shake up the idea of an always and already knowing researcher, to bring in emotional and bodily experiences of academic work, and to help move towards different, more caring, and less conclusive ways of knowledge production. Yet even if we are bound to not (fully) succeed, our collective experiences require us to work towards this goal. With this in mind, we now turn to two phases of fieldwork-based projects which are usually marked by perceived failures: being in the field and 'writing up'.

Doing fieldwork

Fieldwork is never a straightforward application of methodological ruminations and instructions: researchers are unable to control all the factors that determine access to sites and people; they struggle with finding time to do fieldwork; they try to juggle it with teaching responsibilities, administrative duties, personal and family-related considerations; and they depend on ever scarcer funding. In this section, we review three issues that underlie the failures recounted in this volume. We talk about the difficulty of ensuring access and the major role that luck plays in it, the expectation of intimacy, and the still dominant notion of solitary research.

Getting there

In most fieldwork accounts, access problems stay on the margins. Researchers tell us how they established contact, chose their locations, and made their way into the field, but they rarely talk about the obstacles encountered and the affective dimension of the process. The lack of discussion around these issues does not only make researchers fear and suffer failures alone; it also contributes to a limiting vision of fieldwork in which our decisions are based solely on rigorous research design, rather than on the messy relations that we are investigating. Several chapters in this volume speak to these issues. Johannes Gunesch, in his chapter with Amina Nolte, talks about completely changing his Ph.D. research due to security issues: abandoning the planned fieldwork in Egypt, he re-designed his question to study the Egyptian diaspora. Sezer İdil Göğüş notes how she had to start her project anew after the failed coup in Turkey made her focus on AKP activists in Germany instead of doing fieldwork in Turkey.

In the final chapter, Berit Bliesemann de Guevara recounts the difficult start of her research project in Colombia. After initially being granted access to work with political prisoners, a bomb attack in Bogotá for which the prisoners' group claimed responsibility quickly turned them into 'terrorists' and closed them off from research.

Issues of access do not end upon arrival in the field. When confronted with institutional, bureaucratic, and spatial restrictions that can easily derail any outlined project, flexibility and ad-hoc decisions become the norm rather than an exception. This became evident to Holger Niemann at the beginning of his fieldwork at the United Nations headquarters in New York City described in his chapter. From the onset, it was clear that his aim to understand decision-making practices in the Security Council was significantly hindered by the way in which global power relationships were transposed onto the architectural blueprints of the UN headquarters.

As these and other accounts testify, confronted with limited and/or changing access, researchers have to deal with the possibility of the whole project failing, the feeling of inadequacy, and even fears of financial liability.

Empathy, intimacy, and connection

Fieldwork operates with an expectation of empathy – ethnography seeks to humanise the 'other' as a strategy for achieving positive social and political change. In this expectation, intimacy becomes crucial. It is intimacy that is supposed to give us a glimpse of different points of view – we might not *become* the people we study, but by living, thinking, and feeling close to them we should be able to understand how they see the world. In short, we expect to feel 'with and for another'.[4] This quest translates into anticipating personal bonds, solidarity, mutuality, and perhaps even love which are supposed to form during fieldwork. These expectations insidiously set up another trap: as feminist scholars have explored in detail, not only can we never truly feel for another, but sympathy can also easily slip into appropriation (Sylvester 1994; Ferguson 1991). The relationships on which fieldwork depends are complex and fluid – we inevitably fail at intimacy and then deem ourselves incapable of producing valuable ethnographic insights.

In this regard, ethnography failed many of the contributors to this volume, and instead of intimacy they encountered alienation. We feel detached not only from those we knew we would find disagreeable, but, at other times, we feel distance even from those with whom we would expect to build solidarities. In her chapter, Emma Mc Cluskey talks about the decision to abandon her research in a refugee camp after being unable to come to terms with a public whipping that had taken place there. Even though her research was meant to challenge the dehumanisation that happens through EU security practices,

4 We thank Xymena Kurowska for this formulation in her comment on an earlier draft.

she could not generate the empathy that is supposed to underpin this process.

The bonds that we do make are ambiguous and may change quickly. Ewa Maczynska's chapter shows this complexity when her relationship with a research participant becomes almost impossible: in her responses to him, she must take into account his position of a non-European migrant, but this is exactly the position that he wants to escape. Similarly, 'the way forward' can easily come about from mundane misunderstandings. For Lydia C. Cole, deeper understanding of the intersection of politics and psychological care for victims in Bosnia and Herzegovina was made possible by what many would consider a researcher's failure – laughing and crying (inappropriately) in front of her interviewees. Lydia treats these 'failures' as openings – academic, personal, and affective – rather than closures, and shows how 'empathy' can be generated through unexpected means.

Lastly, axes of race, gender, class, and ability intersect to shape our experiences in the field and the relations that make it. Jan Daniel, for example, discusses how gendered norms of what a security researcher should be like shaped his ability to connect to his interlocutors and fuelled feelings of inadequacy. In his conversations with military officials in Lebanon, which he conducted as part of research on interactions between local actors and global organisations, he was perceived as failing to satisfy the ideals of militarised masculinities that were hegemonic in the spaces he conducted his interviews in, a situation that significantly shaped his research. In Sezer İdil Göğüş' case, her position as a Turkish secular woman based at a German institution, yet conducting research on Turkish politics and later the Turkish diaspora, made an imprint on her doctoral project in several ways. Her chapter makes clear that one cannot be sure which of the myriad identity markers condition encounters in the field, in what way, and with what emotional impact. The same is demonstrated by Amina Nolte's research on Israeli infrastructure: at times she was able to use gendered imaginaries of women as harmless to her advantage, yet her Muslim name also brought about complications even before she entered the designated field.

We should note that the discussion of the researchers' identities in this volume has a very specific limitation: all the authors included are racialised as white. This is telling of a larger problem of whiteness that is still at the heart of academia. Although the volume thus misses an important aspect of 'what makes a researcher', we hope that the discussion started here will encourage people to engage in further conversations, including those on race and fieldwork (see e.g. Henderson 2009; Loftsdóttir 2002; Hanson and Richards 2019).

Solitary research and 'having what it takes'

The expectations of integration into the community in the field and of intimacy are closely related to an expectation of being otherwise alone. Even though early ethnographers often relocated with their families who provided both emotional support and research assistance,[5] the fieldworker we usually imagine is a lone hero.[6] During fieldwork, the researcher 'must cut his or her life down' (Van Maanen 2011, 151): not only do we have to suffer through fieldwork as a sort of initiation, but we have to suffer it alone (Ibid., 29) – as if asking for help or working in teams would both detract from our skills and somehow ruin the methodological process.

There are some obvious questions here. Who are those who can afford to 'cut their lives down to the bone'? Caring responsibilities, class and racial locations, and a spate of other factors dictate who is 'allowed' or 'able' to cut themselves off from their 'real life'. And what are the costs of such 'cutting off'? The material and emotional capacities needed to turn a research design into a finished project need to be carefully considered (D'Aoust 2013). As Jakub Záhora's chapter shows via his disclosure of struggles with depression throughout his fieldwork in Israeli settlements and beyond, there are many emotional and personal experiences and qualities that shape what we access, what we do, and how we come to understand our interlocutors.

The current 'epidemic' of mental health issues in academia affecting both staff and students significantly raises the costs of doing research.[7] The fact that many people would endure mental illness, but also plain loneliness and sadness for prolonged periods of times, harks back to the idea that fieldwork is supposed to be somehow challenging, even dangerous – as committed researchers, we should be ready to do anything for data. By talking about our failures to do so, we want to question the ideal of the fieldworker as a disembodied vessel of knowledge smoothly navigating new relationships.

Afterlives of fieldwork

Fieldwork also shapes our expectations of written outcomes – we want others to be equally thrilled about our findings, do justice to the people we spoke to,

5 For example, James Scott's fieldwork for *Weapons of the Weak* (2008) included moving his entire family with him to Sedaka.

6 Different authors discuss the existence of an 'Indiana Jones' image of the fieldworker, yet this image still persists (see Rock 2001, 33; Pachirat 2018, 78–83; Hanson and Richards 2019, 28–29; Clifford 1988).

7 https://www.theguardian.com/education/2019/may/23/higher-education-staff-suffer-epidemic-of-poor-mental-health

and give back to those who gave to us. And we have to publish our findings and secure jobs. We, as many others, have learned the hard way that failures do not stop when leaving the field. There are two issues we want to investigate further in this 'post-fieldwork phase': the expectation of engaged writing and the increasingly neoliberal environment in which early career scholars conduct research.

Fieldwork as textwork

The social sciences are going through a process of raising their 'textual consciousness'. Within IR, discussions of voice, positionality, authorial presence, and reflexivity are employed to fight the fetishisation of objectivity, the soullessness of jargon, and the absence of a positioned author (Doty, 2004). In this context, fieldwork emerges not only as a method, but also as a path towards a different kind of writing. Such expectations set the stage for another failure.

Translating impressions, experiences, and narratives that excite us while in the field into academic texts, judged by their rigour and coherence, often causes deep frustration. Renata Summa addresses this issue in her chapter through a reflection on her struggles to 'capture' her impressions of Sarajevo, where she conducted research on everyday bordering practices. A city that she found mobile, dynamic, and illusive was forced to stand still on paper. This inevitable gap between experience and writing leaves one doubting the value of the text we produce.

In its reflection on the 'textwork' involved in turning fieldwork into academic outputs, we consider this volume to be an intervention. We recognise that both fieldwork and the writing that it inspires requires a constant 'construction and production of self and identity' (Coffey 1999, 1), and the chapters examine these processes through narrative, dialogue, and self-reflection. We thus contribute to the ongoing project of developing novel forms of writing IR and approaching the world differently (Ravecca and Dauphinée 2018; Dauphinée and Inayatullah 2016; Inayatullah 2011). The stories presented are necessarily partial, yet also constitutive of the researchers' selves; by disclosing our failures, the connections between our biographies and our theories are made more accessible to others.

Politics within and beyond our texts

The writing that is born out of (or despite) the failures we recount is engaged in politics: of academia and of the world we study. This engagement is discussed in Katarina Kušić's chapter where she warns that even those sympathetic to efforts to challenge disciplining academic norms should be wary of the proverbial good intentions. In discussing the perils of narrative writing in a neoliberal context, she highlights the need for careful calibration of writing style – something that all the contributions to the volume navigate.

Our texts are also immersed in the power structures they study. In addition to our investments – temporal, emotional, physical – we are in debt to countless people who gave us their time in the faith of us 'doing something' with their input (beyond hard-bound theses sitting in libraries). Desirée Poet's chapter shows how her research with urban indigenous and urban *quilombo* communities in Brazil had to navigate the internal politics of these movements, and wider academia. Her own dissatisfaction with the collaboration that she tried to integrate into her fieldwork shows that the often celebrated 'participatory research' is rarely capable of addressing the power disbalances between the researcher and the researched. Danielle House went through a somewhat similar experience in the course of her project on disappearances in Mexico. Working together with those affected by disappearances enabled her to get closer, and also repay some of the time people invested in her. Desirée's and Danielle's chapters then provide us with glimpses of hope: there might be a way forward – if we take our political commitments seriously and work together for transformations of the structures that inhibit academia as well as what we call the field.

Finally, these concerns are closely related to what determines academic fates today: there is the failure of not publishing, not publishing enough, or not publishing in the right outlets. This type of failure is by now all too familiar – which nonetheless does not diminish its impact on people's fates. The 'hidden injuries of the neoliberal university' (Gill 2010) are not that hidden anymore, but the pressure persists. The imperative to turn our fieldwork into 'successful' publications again points to systemic issues of the academic industrial complex. The current epidemic of 'atypical contracts' – short term, fixed term, or zero hours,[8] exacerbates the publish-or-perish rationality and simultaneously prevents early career scholars from devoting time to follow-up research, writing, and publishing. The 'academic market' pushes early career researchers to enact a never-ending mobility posture (Allmer 2018).[9] As a result, the same intimacy so valued in ethnographic research (and that helps us survive the taxing nature of fieldwork) becomes an obstacle to the competitive self-entrepreneurialism needed to navigate the neoliberal academia.[10]

---

8 In the UK, one third of all academics have fixed-term contracts. See: https://www.hesa.ac.uk/news/24-01-2019/sb253-higher-education-staff-statistics.

9 See https://www.sciencemag.org/careers/2017/08/toll-short-term-contracts.

10 It is noteworthy that even Katherine Verdery (2018, 297), an established American professor of Anthropology, writes that 'the gratifying durability of the connections' in a

Map of the book

The chapters that follow travel from different institutions and fieldwork sites: the failures recounted happened in Palestine, Colombia, New York, Turkey, Brazil, Mexico, Wales, Egypt, Lebanon, Bosnia and Herzegovina, Morocco, and Denmark. What brings them together is their navigating of the juncture between personal experiences and systemic norms. By sharing their experiences and reflections, the authors show that failure is ubiquitous and needs to be dealt with.

The chapters are organised into four sections dealing with different aspects of failure. The first asks 'what makes the researcher'. Chapters by Jan Daniel, Sezer İdil Göğüş, and Jakub Záhora interrogate issues that start well before any actual fieldwork: how we imagine the 'successful researcher' in terms of gender, nationality, and political stance, and how these and other identities and proclivities condition our insights in the field. The second section explores situatedness and ways in which different locations determine what knowledge we can hold. Chapters by Johannes Gunesch and Amina Nolte, Lydia C. Cole, and Holger Neumann show that any reflection on situatedness in the field automatically makes any claim for objective or complete knowledge 'always and already a failure'.

The third section engages the various relations that are forged in the course of fieldwork. Contributions by Emma Mc Cluskey, Desirée Poets, Danielle House, and Ewa Maczynska attend to difficulties in processing the relations on which fieldwork depends and reflect on the impossibility of some of them. In the last part of the volume, Renata Summa and Katarina Kušić deal with the uneasy and often painful process of transforming our embodied experiences, insights, and memories into texts read by others.

The collection concludes with a chapter by Berit Bliesemann de Guevara and Xymena Kurowska that encourages us to reconceptualise failure as 'productive rupture'. They suggest micro strategies – exposure, the capacity for surprise, and reflexivity through positionality – that can help us reinscribe failure collectively and 'problematise the academic frame of mastery'. Although some failures will never be productive – they remain knotted stomachs, shed tears, and discomfort – we want to end with reiterating their (and many of the other authors') call for community, mutuality, and friendship that help mitigate the affective as well as dangerously tangible effects of various failures.

Romanian village where she repeatedly conducted fieldwork over the span of four decades 'stands sharply opposed to [her] life at home, fractured by multiple moves from place to place'.

We prepare this volume in a peculiar time of heightened pressures of success and the political and analytical potentials of failure. With Halberstam (2011, 2–3), we remain convinced that 'failing, losing, forgetting, unmaking, undoing, unbecoming, not knowing may in fact offer more creative, more cooperative, more surprising ways of being in the world'. We treat failure as both a productive opening *and* as a resistance to the imperative of production. It is in this spirit that we invite you to explore the experiences that follow – experiences that are always singular, unique and individual, yet speak to shared concerns of those who set out to look for the international in the field.

* *Both authors would like to thank Xymena Kurowska and Berit Bliesemann de Guevara for their comments on earlier versions of this chapter. The support of the Economic and Social Research Council (UK) (ES/T009004/1) for a part of this research is gratefully acknowledged by Katarina Kušić. Jakub Záhora's work on this volume was supported by the Charles University Research Programme 'Progres' Q18 – Social Sciences: From Multidisciplinarity to Interdisciplinarity.*

References

Ackerly, Brooke A., Maria Stern, and Jacqui True, eds. 2006. *Feminist Methodologies for International Relations*. Cambridge; New York: Cambridge University Press.

Allmer, Thomas. 2018. 'Precarious, Always-on and Flexible: A Case Study of Academics as Information Workers'. *European Journal of Communication* 33 (4): 381–95.

Aradau, Claudia, Jef Huysmans, Andrew Neal, and Nadine Voelkner, eds. 2015. *Critical security methods: New frameworks for analysis*. Harvard: Routledge.

Aradau, Claudia, and Jef Huysmans. 2014. 'Critical Methods in International Relations: The Politics of Techniques, Devices and Acts'. *European Journal of International Relations* 20 (3): 596–619.

Bliesemann de Guevara, Berit, and Morten Bøås, eds. 2020. *Doing Fieldwork in Areas of International Intervention: A Guide to Research in Violent and Closed Contexts*. Bristol: Bristol University Press.

Cerwonka, Allaine, and Liisa H. Malkki. 2007. *Improvising Theory: Process and Temporality in Ethnographic Fieldwork*. Chicago: University of Chicago Press.

Clifford, James. 1988. 'On Ethnographic Self-Fashioning: Conrad and Malinowski'. In *The Predicament of Culture: Twentieth-Century Ethnography, Literature, and Art*, 92–113. Cambridge: Harvard University Press.

Clifford, James, and George E. Marcus, eds. 1986. *Writing Culture: The Poetics and Politics of Ethnography*. Berkeley; Los Angeles: University of California Press.

Coffey, Amanda. 1999. *The Ethnographic Self: Fieldwork and the Representation of Identity*. London: SAGE.

Daigle, Megan. 2016. 'Writing the Lives of Others: Storytelling and International Politics'. *Millennium: Journal of International Studies* 45 (1): 25–42.

D'Aoust, Anne-Marie. 2013. 'Do You Have What It Takes? Accounting for Emotional and Material Capacities'. In *Research Methods in Critical Security Studies: An Introduction*, edited by Mark B. Salter and Can E. Mutlu, 33–36. London; New York: Routledge.

Dauphinée, Elizabeth. 2007. *The Ethics of Researching War: Looking for Bosnia*. Manchester; New York: Manchester Univ. Press.

———. 2016. 'Narrative Engagement and the Creative Practices of International Relations'. In *Reflexivity and International Relations: Positionality, Critique and Practice*, edited by Jack L. Amoureux and Brent J. Steele, 44–60. London; New York: Routledge.

Dauphinée, Elizabeth, and Naeem Inayatullah, eds. 2016. *Narrative Global Politics: Theory, History and the Personal in International Relations*. London; New York: Routledge.

Doty, Roxanne Lynn. 2004. 'Maladies of Our Souls: Identity and Voice in the Writing of Academic International Relations'. *Cambridge Review of International Affairs* 17 (2): 377–92.

Ferguson, Kathy E. 1991. 'Interpretation and Genealogy in Feminism'. *Signs* 16 (2): 322–39.

Gill, Stephen. 2010. 'Breaking the Silence: The Hidden Injuries of The Neoliberal University'. In *Secrecy and Silence in the Research Process:*

Feminist Reflections, edited by Róisín Ryan-Flood and Rosalind Clair Gill, 228–44. London; New York: Routledge.

Glasius, Marlies, Meta De Lange, Jos Bartman, Emanuela Dalmasso, Aofei Lv, Adele Del Sordi, Marcus Michaelsen, and Kris Ruijgrok. 2018. *Research, Ethics and Risk in the Authoritarian Field*. Cham, Switzerland: Palgrave Macmillan.

Halberstam, Judith. 2011. *The Queer Art of Failure*. Durham: Duke University Press.

Hanson, Rebecca, and Patricia Richards. 2019. *Harassed: Gender, Bodies, and Ethnographic Research*. Oakland: University of California Press.

Haraway, Donna. 1988. 'Situated Knowledges: The Science Question in Feminism and the Privilege of Partial Perspective'. *Feminist Studies* 14 (3): 575–599.

Harding, Sandra. 1992a. 'After the Neutrality Ideal: Science, Politics, and "Strong Objectivity"'. *Social Research* 59 (3): 567–587.

———. 1992b. 'Rethinking Standpoint Epistemology: What Is" Strong Objectivity?"'. *The Centennial Review* 36 (3): 437–470.

Henderson, Frances B. 2009. '"We Thought You Would Be White": Race and Gender in Fieldwork'. *PS: Political Science & Politics* 42 (2): 291–294.

Inayatullah, Naeem, ed. 2011. *Autobiographical International Relations: I, IR*. London: Routledge.

Jemielniak, Dariusz, and Monika Kostera. 2010. 'Narratives of Irony and Failure in Ethnographic Work'. *Canadian Journal of Administrative Sciences / Revue Canadienne Des Sciences de l'Administration* 27 (4): 335–47.

Kapiszewski, Diana, Lauren M. MacLean, and Benjamin Lelan Read. 2015. *Field Research in Political Science: Practices and Principles*. Cambridge: Cambridge University Press.

Kurowska, Xymena. 2019. 'When One Door Closes, Another One Opens? The Ways and Byways of Denied Access, or a Central European Liberal in Fieldwork Failure'. *Journal of Narrative Politics* 5 (2): 71–85.

Law, John. 2007. *After Method: Mess in Social Science Research*. London: Routledge.

Lie, Jon Harald Sande. 2013. 'Challenging Anthropology: Anthropological Reflections on the Ethnographic Turn in International Relations'. *Millennium: Journal of International Studies* 41 (2): 201–20.

Loftsdóttir, Kristín. 2002. 'Never forgetting? Gender and racial-ethnic identity during fieldwork'. *Social Anthropology* 10 (3): 303–317.

Marcus, George E., and Michael M. J. Fischer. 1986. *Anthropology as Cultural Critique. An Experimental Moment in the Human Sciences*. Chicago and London: Chicago University Press.

Naumes, Sarah. 2015. 'Is All "I" IR?' *Millennium: Journal of International Studies* 43 (3): 820–32.

Nordstrom, Carolyn, and Antonius C. G. M. Robben, eds. 1995. *Fieldwork under Fire: Contemporary Studies of Violence and Survival*. Berkeley: University of California Press.

Pachirat, Timothy. 2018. *Among Wolves: Ethnography and the Immersive Study of Power*. New York: Routledge.

Page, Tiffany. 2017. 'Vulnerable Writing as a Feminist Methodological Practice'. *Feminist Review* 115 (1): 13–29.

Rabinow, Paul. 1977. *Reflections on Fieldwork in Morocco*. Berkeley: University of California Press.

Ravecca, Paulo, and Elizabeth Dauphinée. 2018. 'Narrative and the Possibilities for Scholarship'. *International Political Sociology* 12 (2): 125–38.

Rock, Paul. 2001. 'Symbolic Interactionism and Ethnography'. In *Handbook of Ethnography*, edited by Paul Atkinson, Amanda Coffey, Sara Delamont, John Lofland, and Lyn Lofland, 26–38. London: SAGE.

Salter, Mark B., and Can E. Mutlu, eds. 2012. *Research Methods in Critical Security Studies: An Introduction*. New York: Routledge.

Schatz, Edward, ed. 2009. *Political Ethnography: What Immersion Contributes to the Study of Power*. Chicago; London: The University of Chicago Press.

Scott, James C. 2008. *Weapons of the Weak: Everyday Forms of Peasant Resistance*. New Haven; London: Yale University Press.

Sjoberg, Laura. 2019. 'Failure and Critique in Critical Security Studies'. *Security Dialogue* 50 (1): 77–94.

Smith, Robin James, and Sara Delamont, eds. 2019. *The Lost Ethnographies: Methodological Insights from Projects That Never Were*. Bingley, UK: Emerald Publishing.

Steele, Brent J., Harry D. Gould, and Oliver Kessler, eds. 2019. *Tactical Constructivism, Method, and International Relations*. Abingdon; New York: Routledge.

Stefan, Melanie. 2010. 'A CV of Failures'. *Nature* 468 (7322): 467–467.

Sylvester, Christine. 1994. 'Empathetic Cooperation: A Feminist Method For IR'. *Millennium: Journal of International Studies* 23 (2): 315–34.

Thomson, Susan, An Ansoms, and Jude Murison, eds. 2013. *Emotional and Ethical Challenges for Field Research in Africa*. London: Palgrave Macmillan UK.

Van Maanen, John. 2011. *Tales of the Field*. Chicago: University of Chicago Press.

Verdery, Katherine. 2018. *My Life as a Spy: Investigations in a Secret Police File*. Durham: Duke University Press.

Vrasti, W. 2008. 'The Strange Case of Ethnography and International Relations'. *Millennium: Journal of International Studies* 37 (2): 279–301.

Wibben, Annick T. R., ed. 2016. *Researching War: Feminist Methods, Ethics and Politics*. London; New York: Routledge.

Yanow, Dvora, and Peregrine Schwartz-Shea, eds. 2006. *Interpretation and Method: Empirical Research Methods and the Interpretive Turn*. Armonk, N.Y: M.E. Sharpe.

Part I

Successfully Making the Researcher

1

Fieldwork, Feelings and Failure to Be A (Proper) Security Researcher

JAN DANIEL

It is ten minutes past eight in the morning and I have finally arrived at a military base at the outskirts of Vicenza – the place where I am supposed to meet and interview Major Pierpaolo, a high-ranking officer in a research department of an international military training institute. I am late, it is raining, and I made a bad decision to walk through the town, instead of taking a bus, so I am wet and sweating. When my interlocutor, a cleanly shaven, large man dressed in a fitted dark blue uniform of Italian *carabinieri*, arrives to pick me up at the main gate, I can immediately sense that he was expecting a different person. Perhaps more senior, perhaps dressed in something else than jeans and a coloured shirt, or perhaps even someone with a military background. In the end, I am supposed to be a representative of a governmental research institute (or so says my affiliation) and I am doing research on serious military issues. As he walks me from the gate to his office and I unsuccessfully try to start a conversation, I start to think that perhaps I am not the right person for this research. This is not what I expected fieldwork to be like.

He gives me a tour of the facility and grudgingly answers some of my questions related to my research project, while indicating that he has better things to do than talk with a young, nervous, and visibly non-military guest. A fleeting sense of shared understanding among us is established only as we watch a group of non-European peacekeepers trying to perform a mock raid of a locked building and grotesquely failing to uphold a proper formation and ram through the doors. He rolls his eyes, gives me a slightly apologetic look and mutters that learning is a process. However, he quickly regains his detached and disinterested way of interacting with a young civilian dressed in overtly casual clothing and with a visibly non-military posture. As he walks me through the training grounds back to his office, I start to think that even though I got some interesting 'data' for my research project, this encounter feels like a failure.

I now know these moments and feelings that come with them quite well. The atmosphere during the interview and throughout the whole day was very formal and cold at best. The sense of closeness and mutual interest that sometimes appears during such research encounters was not there. I felt that I was not a conversation partner but rather an unwelcomed nuisance, a young civilian without military experience or a clue about 'real' military life. I have experienced similar situations before and I also know it is not something unusual. Many research encounters are deeply unsatisfactory for both sides and these failures happen for a myriad of different reasons. Still, the feelings which emerge during these moments are everything but pleasant. Among the dominant ones are an overwhelming sense of despair about potentially ruining a research project, anger that I am unable to establish a proper working relationship with my interlocutor, and self-doubt stemming from the question of whether I am able to conduct any field research at all.

Feelings of failure

Following the editors' call to reflect on the notions of fieldwork and failure, I inquire into my personal feeling of failure during fieldwork and conditions that contributed to it. Countless fieldwork manuals for junior researchers explicitly state that the research conducted with 'real people in real places' is a stressful and unpredictable endeavour and it takes an emotional toll on the researcher. Many also mention the importance of 'impression management' needed to 'fit' the researcher with the studied group and bring him or her closer to the researched individuals to gain their trust, recognition and maintain access (e.g. Hammersley and Atkinson 2007, 66–71). These issues get even more pronounced in the cases of 'studying up' and research done on powerful actors and/or security professionals, where access is difficult and lack of trust towards researchers implicit (Baker et al. 2016; Ben-Ari and Levy 2014; Gusterson 1997; Kuus 2013). As interview opportunities are granted only rarely and the refusal of access can lead to the failure of a whole research project (see e.g. Kurowska, 2019), the perceived costs of potential failure and resulting pressure on getting the interview right could be felt as quite high.

This text presents an attempt to reflect on personal feelings of failure to conduct field research in the settings dominated by men of power – primarily military personnel and governmental security bureaucrats. I approach the notion of failure through a set of feelings – inner emotional states translated (and translatable) into words (Hutchison and Bleiker 2014, 501) – that were produced by my failed and failing research encounters. I am fully aware of an inevitable distortion of my memories related to the described events and emotions which accompanied them (as some moments have been unintentionally blended with others in the narrative reconstruction). Nevertheless, I am also convinced that these emotions, affects, and feelings should be productively interrogated to uncover the wider structural conditions which formed both the researcher and the idea of fieldwork – or in other words, the baggage that we bring with us to the particular moment of the research encounter (Åhäll 2018, 40; Davies and Spencer 2010, 23).

In the following paragraphs, I briefly trace some origins of this baggage as well as particular contexts of my feelings of failure experienced during fieldwork. I believe that some of my experiences might resonate with those of other researchers and stimulate their own reflections, however this exercise is also a personal attempt to think through some of the moments which I remember for their impact on my future research strategies or for their intensity. Some of them are related to my feeling of inability to establish a productive rapport with my interlocutors and resulting feelings related to personal inadequacy; others emerged from the messy nature of a fieldwork process, the sense of failing at it and my reactions to these failures. These feelings are by far not as traumatic (or dramatic) as those experienced by researchers working in violent environments (e.g. Al-Masri 2017; Monaghan 2006; Nordstrom and Robben 1995; Woon 2013). In fact, compared to them, they are admittedly quite banal. Nevertheless, they point to the embodied nature of field research, where the researcher faces his or her 'inescapable corporeality and emotional vicissitudes' (Monaghan 2006, 226; see also Coffey 1999; Vanderbeck 2005) – corporeality and emotions that inevitably influence mutual positioning of both researcher and his or her interviewees during their encounters.

Approaching these topics through the instances of my feelings of personal failure, I focus in particular on three main issues – the importance of already existing expectations as a benchmark against which failure is assessed, the intimate nature of fieldwork as an activity that inserts a researcher into particular relationships with his or her informants, and finally, the transformation of particular feelings in time.

Fieldwork as an adventure

Only recently I realised how much the feelings of failure I experienced during certain periods of my research were influenced by my undergraduate studies. I was trained in a security studies section of a political science department where fieldwork and related direct exposure to the studied issues were highly valued. According to stories circulated within classrooms, my (predominantly male) instructors rubbed shoulders with private military contractors, members of Shia militias, or Chechen rebel fighters – and they frequently spoke about their experiences in the associated risky research terrains. A particularly popular story shared in the methodology class on qualitative research involved one of the assistant professors and his participant observation among local right-wing skinheads; the story included his experience of drinking beer and singing while narrowly avoiding a bar fight. Another recounted a meeting with members of Yemeni tribes that ended in a shooting competition and an invitation to practice firing an RPG.

I did not look up to my professors, but through their lectures and stories and the literature I was assigned to read, I developed an idea of fieldwork, particularly in a wider area of security studies, as a dangerous adventure which was rewarded by first-hand access to the researched groups and a certain camaraderie with their members. Needless to say, potential failure was never mentioned and if it was, it was only for a comical effect. Similarly, any potentially discomforting emotions, such as fear or anxiety, were left out of a story or mentioned merely as a passing temporary distraction that can complicate the pursuit of research. The figure of a researcher conducting fieldwork in such narratives corresponded to a masculine hero who bravely and rationally faces the difficulties he encounters and returns from the field with first-hand knowledge of the studied issues (for further reflection on reproduction of hegemonic masculine values and fieldwork see Vanderbeck 2005).

In sum, the fieldwork, as I was taught to imagine it in my methodological classes and through stories told by my instructors, was not without its difficulties and potential failures. However, these were primarily of a physical nature that came with 'dangerous' settings where the research was taking place. On the other hand, fieldwork was not supposed to be emotionally demanding. It rather corresponded to the stories of adventure and exciting encounters with people and places about which I, at that time, only read in books and heard in the media.

Failures and encounters

My first fieldwork, conducted during the second year of my Ph.D. studies, felt very different. A director and a deputy director of one of the smaller Czech security agencies, who were sitting behind a large wooden table in an office in the centre of Prague, started our discussion by reversing the roles of the interviewer(s) and interviewee(s) and examining the depth of my knowledge of the studied issue (on similar experience see also Kapiszewski et al. 2015, 86). When I somehow mumbled more or less satisfactory answers, they continued the interview by voicing their disdain for political science and its lack of any useful insights as well as for political scientists researching subjects they do not have any practical experience with. The shivering, nervous, and perplexed sound of my voice on the interview recording manifested that this was among the more stressful encounters that I experienced, and it left a strong mark in my memory. 'Looking forward to seeing you next time, dear student', they said as I was preparing to leave, articulating clearly the relationship between us and leaving me relieved, but also embarrassed and slightly angry about being so decisively put in my (supposed) place. In retrospect, the interview was not a complete failure as the officers started to engage with my questions after the initial clarification of hierarchies and they later continued to cooperate with me and my colleague on other projects, but looking back, it was indicative of future feelings of failure.

A couple of months later I embarked on my first 'proper' fieldwork. My project concentrated on local practices of UN peacekeeping in southern Lebanon and relations between peacekeepers, local civilian communities, and political actors. It did not go as expected. The interesting (and dangerous) people I wanted to interview did not want to speak with me as they were not authorised or interested to speak with a foreign researcher. The access to the main site was complicated by endless and tiresome bureaucratic procedures – it once caused me to be returned from a checkpoint leading to my studied area and missing an important interview that took me nearly a month to arrange. Moreover, the data which I gleaned from other sources did not conform with the concepts and theories which my Ph.D. project was based on. The dominant feeling at that time was not one of excitement but rather one of frustration with and anxiety over where my research was heading.

The sense of overall failure was only strengthened during a consultation with a prominent US journalist based in the country, a large bearded man in his fifties. He covered security issues in the region I studied for several US newspapers and magazines, and his articles at that time formed a significant part of my thinking about my research. He also brushed off most of my inputs into the conversation. My questions about inner workings of the local security field were met with a shrug and my ideas about potential gatekeepers who can help me to gain access to the studied area were dismissed without a feasible alternative. I felt intense embarrassment and even humiliation stemming from his reactions and general lack of interest. His concluding words 'OK, time for another one. Yeah, and thank you for the tea', which he uttered as he moved to another table with a waiting young guy and leaving me the bill to pay (a small detail which made me particularly upset at that moment), just added to the overall feeling of frustration and rejection. I felt that I not only failed to gain any meaningful insight into the studied issues, but also that I was failing in doing fieldwork as I imagined it – I could not establish collegial or even friendly relations with my interlocutors.

Yet, the feelings of failure can also change, sometimes quite abruptly. In fact, two encounters which followed the most intensely felt failures were also among the most satisfactory ones out of those I conducted during the initial stages of my doctoral research. One of them took place when my shared taxi took a detour and then got lost on a rainy evening in the eastern hilly outskirts of Beirut, where I was supposed to meet a high-ranking UN officer in a compound of one of the many UN agencies. I arrived at the meeting place more than an hour late, at the time when my interview was supposed to finish. However, my visible despair over not being able to carry-on my research in a professional manner changed my initially reserved and irritated interlocutor into a more open and welcoming person. In the end, the welcoming reaction of my interlocutor to my initial failure helped to produce a particularly insightful informal conversation on the state of the country, its infrastructure, and the role of the UN. What felt like a horrible failure caused by my lack of planning was in a few minutes transformed into a pleasant experience of meeting a welcoming and helpful person who was, given his own 'baggage', able to relate to the everyday difficulties of carrying out fieldwork.

A similar shift in my personal feeling of failure happened some time later, when after two months of waiting I was finally granted a permission to interview peacekeepers at the Lebanese-Israeli border. Experiencing a difficult personal period that added to my overall state of desperation with my fieldwork, I lost my voice due to a sore throat and staying out late on a cold night. Barely recovered, I travelled to the border region to meet my interviewees, only to lose my voice again in the morning before the interviews even started. Until now, it is difficult for me to think of a better example of a particularly deep feeling of complete failure. The question of how I could be so stupid and lose my voice, the only thing that I, in the end, need to perform my interviews, kept popping up in my head when I struggled to produce basic sounds resembling some words and introduce myself to an Indian peacekeeper, a young commander of a military-community outreach unit, and his deputy. Seeing my condition, my interlocutors reacted by taking me to a canteen in their compound and provided me with herbal tea and some medicine. I slowly regained my voice and, while I kept losing it throughout the whole day, we managed to talk. As the deputy-commander walked me out of the base in the afternoon, he asked me an unexpected question about my age. 'Good, we are the same age. We can be friends', he responded when hearing my answer. As with the previous encounter, the sense of desperation, my failure to behave as a 'normal' researcher, the strangeness of the situation for both sides, and the willingness of my interlocutors to react to it in an open and caring way transformed the atmosphere of the meeting. My emotional reaction to it as well as our relation had enabled us to establish a different, potentially more productive, form of rapport – in the case of the Indian peacekeeper, a rapport which even developed into a certain kind of friendship maintained through periodic updates on a messaging app.

Nevertheless, fieldwork encounters are unpredictable and not all failures turn into something more pleasant. And even if one manages to gradually enact some distance between themselves and the mishaps in the field, they still have an impact. A few years after the story recounted in the introduction, I found myself sitting behind a heavy wooden table in an office decorated with old rifles and memorial plaques from NATO training exercises. 'Do you even know what the peacekeeping operations are?', a high-ranking official at the Czech Ministry of Defence asked me. 'Do you know who General Dallaire was? Do you know what happened in Rwanda?' Of course, I did. However, that was not enough for him. 'So why are *you two* writing a report on the Czech involvement in the UN peacekeeping operations? What do you even know about the Army logistics and training practices?' At this point, it became obvious that my research partner and I were not the people he expected when he (or rather his assistant) warmly answered our email asking for an interview. Perhaps he expected someone more senior, or perhaps even someone from the military and with military experience. After further queries from him, which took up all the time for the interview, we were told that our time was up, and that we could send further questions by email. We shared a feeling of despair for not managing to productively conduct an important interview, anger for not even being given a chance to try to do so, and anxiety over the future of the project. Though it would be uplifting to conclude the paper with a story of turning a failure into a sort of success, my experience of doing research among diplomats, security bureaucrats, military officials and other 'men of power' produced probably more stories of failure like this and the one which opened the paper (see also Baker et al. 2016) than stories about the unexpected turning of failure into success.

Conclusion

In conclusion, I briefly return to the three aspects of fieldwork failure and the connected feelings. To say that failure does not make sense without certain expectations and connected normative standards is to state something obvious. Many of my initial feelings of failure during my own fieldwork stemmed from unrealistic expectations about adventurous, controlled and masculine fieldwork that I developed in my early studies. Fieldwork, as I came to know it during my Ph.D. studies is, however, not like that. Fieldwork is, or could be, among many other things, messy, and deeply frustrating and failure is unavoidable. Many stories of failure described above stem from various accidents and contingencies that make fieldwork often a very unpredictable experience over which a researcher has only limited control. This is not to say that planning fieldwork is impossible, irrelevant, or that better planning would not limit the impact of certain failures. This simple advice is emphasised by many fieldwork manuals (e.g. Kapiszewski et al. 2015). However, knowing before embarking on my first longer bout of fieldwork that the moments of failure and the feelings they produce are shared by many researchers might have made certain moments a bit more bearable.

I also tried to show through the text that my feelings of failure have often emerged from unsatisfactory relations with my interlocutors and my (perceived) failing to fit with their expectations of what a proper security researcher looks and behaves like. In a way this also speaks to the contingency of fieldwork as some such failures can destabilise the roles of a researcher and an informant and produce a different, potentially even closer, form of rapport, while others lead to outright rejection. However, beyond contingency, the feelings that I engaged with in this paper are also inseparable from the very nature of the research encounter as a meeting of two (or more) people with their own 'baggage' of previous experiences, expectations, and emotional investments in the given situation.

These feelings stem from a specific understanding of relations between the researcher and the researched and the unrealistic expectation of a certain closeness between the two.[1] In other words, our interlocutors are not (automatically) our friends and we should not expect them to be, as our roles in this type of encounter are different. There is an instrumental interest on both sides: I want to learn certain information and my interlocutors want to tell certain stories and/or are curious about the experience of being interviewed. Taking this into account might help to separate oneself from some unpleasant moments which happen during fieldwork and limit the potential emotional damage – something which I have been thinking about since the experiences discussed here, and which I have yet to learn how to fully apply in practice.

Finally, failure, if approached through the feelings, emotions, and affects connected to it, has its specific afterlife. There are many contradictory feelings which I have experienced during my interviews and fieldwork for different projects. In retrospect, many of the failures recounted above and the feelings associated with them could serve as interesting data. I can use them

[1] I would like to thank a discussant of an early version of this paper, Xymena Kurowska, for pointing this out.

to grasp a certain form of relationship and identity-performance which would contextualise the given situation and help me understand more about the social and organisational settings in which my interlocutors are embedded. In fact, this whole text is a result of such reflection. However, the comfort of a detached position is not present during moments when failure to establish a productive and mutually respectful relation is felt. As much as I deeply enjoy doing fieldwork, these moments often make me feel like I am a failure.

References

Åhäll, Linda. 2018. 'Affect as Methodology: Feminism and the Politics of Emotion'. *International Political Sociology* 12 (1): 36–52.

Al-Masri, Muzna. 2017. 'Sensory Reverberations: Rethinking the Temporal and Experiential Boundaries of War Ethnography'. *Contemporary Levant* 2 (1): 37–48.

Baker, Catherine, Victoria Basham, Sarah Bulmer, Harriet Gray, and Alexandra Hyde. 2016. 'Encounters with the Military: Towards a Feminist Ethics of Critique?' *International Feminist Journal of Politics* 18 (1): 140–154.

Ben-Ari, Eyal, and Yagil Levy. 2014. 'Getting Access to the Field: Insider/Outsider Perspectives'. In *Routledge Handbook of Research Methods in Military Studies*, edited by Joseph Soeters, Patricia M. Shields, and Sebastiaan Rietjens, 9–18. London; New York: Routledge.

Coffey, Amanda. 1999. The *Ethnographic Self: Fieldwork and the Representation of Identity*. London: SAGE Publications.

Davies, James, and Dimitrina Spencer, eds. 2010. *Emotions in the Field: The Psychology and Anthropology of Fieldwork Experience*. Stanford: Stanford University Press.

Gusterson, Hugh. 1997. 'Studying Up Revisited'. *PoLAR: Political and Legal Anthropology Review* 20 (1): 114–119.

Hammersley, Martyn, and Paul Atkinson. 2007. *Ethnography: Principles in Practice*. London and New York: Routledge.

Hutchison, Emma, and Roland Bleiker. 2014. 'Theorizing Emotions in World Politics'. *International Theory* 6 (3): 491–514.

Kapiszewski, Diana, Lauren M MacLean, and Benjamin Lelan Read. 2015. *Field Research in Political Science Practices and Principles*. Cambridge: Cambridge University Press.

Kuus, Merje. 2013. 'Foreign Policy and Ethnography: A Sceptical Intervention'. *Geopolitics* 18 (1): 115–131.

Monaghan, Lee F. 2006. 'Fieldwork and the Body: Reflections on an Embodied Ethnography'. In *The SAGE Handbook of Fieldwork*, edited by Dick Hobbs and Richard Wright, 226–241. London: SAGE Publications.

Nordstrom, Carolyn, and Antonius C. G. M. Robben, eds. 1995. *Fieldwork under Fire: Contemporary Studies of Violence and Survival*. Berkeley: University of California Press.

Vanderbeck, Robert M. 2005. 'Masculinities and Fieldwork: Widening the Discussion'. *Gender, Place & Culture* 12(4): 387–402.

Woon, Chih Yuan. 2013. 'For "Emotional Fieldwork" in Critical Geopolitical Research on Violence and Terrorism'. *Political Geography* 33 (March): 31–41.

2

Negotiations in the Field: Citizenship, Political Belonging and Appearance

SEZER İDİL GÖĞÜŞ

In ethnographic research, we always talk about fieldwork, our conduct and the encounters in the field. Coffey states that fieldwork has 'identity dimensions' (Coffey 1999, 1) and it is personal. Because the researchers are 'human beings with specific histories, capacities, and characteristics' (Schwartz-Shea and Yanow 2012, 67), their identities have an impact on the research (Lavis 2010; Chege 2015). Fieldwork is also 'social and relational' (Hume and Mulcock 2004, xxii; also see Gupta and Ferguson 1997). In that sense, it is a two-way sense-making (Chege 2015) process, in which researchers and interlocutors try to comprehend each other: researchers aim to grasp the realities in the field, and the interlocutors attempt to understand who we are, our actions and interests (Chege 2015). In other words, a negotiation occurs between the researcher and the researched, in which both sides would try to estimate how much and in what way they will/could fit in with others (Chatman et al. 2005). Katherine Verdery states that such interactions can cause a reciprocal identity creation (Verdery 2018, 23). Hence, new identities can be assigned, created and formed for the researcher by the interlocutors and/or by themselves (Lavis 2010, 317).

During my fieldwork in Turkey in 2016[1], I also had to face several negotiations. I spent two months (April and May) in several cities and districts in Turkey (Istanbul, Ankara, Muğla and Rize) and conducted explorative field research focusing on political socialisation practices amongst members of the

---

[1] Due to the political situation after the June 2016 attempted coup in Turkey, I had to change my dissertation project. Currently, I am working on Turkish diaspora politics and political subjectivities in Germany.

ruling party – the Justice and Development Party (AKP). I aimed to analyse and understand the perspectives of the party members and their reasons for supporting the party in spite of growing criticism of its politics, both at home and abroad. I visited party offices, participated in party events and conducted interviews with members. Although I was aware of possible challenges I might face in the field due to the political situation in Turkey, I did not expect major difficulties, because I was doing ethnography at 'home'. However, I noticed tensions between my interlocutors and me based on various assumptions regarding my person and my interest. My place of residence, my interest in learning their perspectives, and even my eyeglasses both hinted at parts of my identity and created new versions of me.[2] Therefore, in this chapter, by documenting stories from my fieldwork, I aim to reflect on challenging incidents, which represent moments of negotiations in the field.

Native or foreign?

During my fieldwork, I visited one of the small holiday towns in Muğla. The town is overcrowded in the summer but becomes deserted in the winter. When I visited in April, the tourist season had not started yet, and the town was still quite empty.

There, I had the chance to talk not only with women but also with men who are active in party politics.[3] They were eating lunch, drinking tea and coffee, or discussing politics in the common area. On the first day I spent at the office, I talked with several people of various ages: a young woman in her early 20s; women in their late 40s and 50s; and men in their 50s and 60s. Reaching various members of the party was uncomplicated: while I was sitting in the room, different faces came by, commented on a subject of discussion and then left the room. Many of them were quite open to explaining their views on Turkish politics, the AKP and why they chose to work for the party. While the atmosphere was relaxed and people were generally happy to speak with me, I was also asked by some party members to show an official document from my university to prove that I was really writing a dissertation. Many of them also did not allow me to record our conversations.

When I explained that I was associated with a German university, a few people pointed at a retired couple from Germany, suggesting that I should talk with them. The next day, I went to the office again to meet with this couple. They told me that they had lived for some years in Germany and possess

[2] For more on the multiplication of identity in the field see Verdery 2018.

[3] Interestingly, in other offices, I was sent directly to the women's branch of the party, and I could mostly talk with women who worked in the office.

both German and Turkish citizenship, but are currently living in Turkey. When I asked their permission to carry out an interview, and asked whether I would be allowed to use a recording device, he [the husband] did not want to see the official document from the university and allowed me to record our talk.

When I started to work on the data I had collected, I realised something else in my field notes on the elderly man: I saw myself as somebody from Germany, not as someone living temporarily in Germany. Regarding this situation, I commented in my journal: 'He [the husband] was quite interested in my research and asked me if I could send him my thesis afterwards because he would like to learn from my results […] Interestingly, he was using German words in his sentences. I think it was some sort of reference to his background in Germany or maybe he thought I would understand him better. He might have found me likeable because I am *from* Germany' (author's fieldnotes, 22 April 2016).

This realisation about how I defined myself made me think about what it means to be a 'native', a 'halfie' and a 'foreign' researcher, and/or what the advantages and disadvantages are of each. Ohnuki-Tierney states that native anthropologists have a 'more advantageous position in understanding the emotive dimensions of behaviour' (Ohnuki-Tierney 1984, 584). Indeed, native anthropologists might have intimate or in-depth knowledge of the interlocutors' daily routines and are likely to be familiar with their culture. Similar to being native, being a halfie – or bicultural – can also imply an insider perspective as halfies or biculturals can position themselves in two communities (or maybe even more) (Abu-Lughod 1991). Halfie/bicultural researchers may also face representation issues in the field regarding the 'self' and the 'other'[4] during interactions with the interlocutors: 'Are you a native or a foreigner?'

In my case, what was surprising was the realisation of the shift from being native to becoming bicultural. Before I stepped into the field, I had assumed that I was a researcher from Turkey who was currently residing and working in Germany and who, at the same time, had an insider perspective. But through interactions with the people in the field and due to their perceptions of me, I became a hybrid: a Turkish-German. Following that, I received explanations from my interlocutors about Turkish history and the country. Interestingly, my interlocutors' perceptions of me were echoed subconsciously in my field notes, as noted above. Even though it was a discovery of my identity, it also made me concerned about whether I had lost the insider perspective on the country and its people.

[4] See the 'Writing Culture' debate on how 'the other' is represented by the researcher: Clifford & Marcus (1986) and Abu-Lughod (1991).

Some kind of team member?

The head of the AKP's women's branch in another small town in Muğla invited me to an informal coffee meeting with the other active female members of the party. In this small group setting, I was able to grasp their views on the party and why they think their 'public service can only do well for Turkey's future' (interview with Women's Branch Muğla, 28 April 2016). After two hours of talking, and when I thought it was time to leave, I received a request from one of the women there: they wanted to take a photo with me and put it online on the party webpage. I sensed that they were quite sure about my support for the AKP and believed that a possible future collaboration would also be of interest to me – I had approached them and shown interest in getting to know them so I must, therefore, be a party supporter.

Being photographed was a big dilemma for me: firstly, I did not want to turn down their request and risk breaking the trust I was trying to build; secondly, I did not want to have any kind of connection to the party, which I do not support but rather criticise. I refused the request and argued vaguely that it might harm my impartiality as a researcher. That was the turning point for our relationship. I wondered whether they were disappointed and/or if they started to see me not as one of them but as some 'other'. In my journal, I noted: 'They also wanted to know my view on the issues about which we were talking: on the AKP, the future of Turkey, the success of the AKP'. The head of the women's branch, Songül, said, 'It is now our turn; we will ask you questions'.[5] They wanted to hear more about my own political views and my family's political background. I realised that being from Turkey and doing research in Turkey puts me in a complicated position: 'They were curious about my background and whether they could relate to me' (author's fieldnotes, 27 April 2016). This situation might have been similar to that of a foreign researcher; they might have to talk about themselves and their political views as well. However, as a 'native' researcher, I could not avoid giving a proper answer to some questions, such as which party I voted for. My interlocutors were familiar with the political history and the polarisations in the country, and I was as well. Thus, I was afraid that my answer might damage the relationship: if I told them the truth about my political views, I might not have the chance to get 'deep hanging out' (Geertz 1998) with them. At the same time, I was also concerned that not revealing my honest views to them might create ethical issues. In hindsight, I recognise that my vague answer might have been equally unsatisfying. Also, my actions should not have been so influenced by my fear of failure and of losing my connection with them because it is indeed in the nature of fieldwork that you sometimes build trust and in other cases you lose the contact.

5 In order to protect the anonymity of my interview partners, I have changed their names.

It is mainly discussed that ethnographic research should be conducted from the position of 'some kind of team member' (Reiter-Theil 2004, 23; cited in Lewis and Russell 2011, 400). However, during such highly politically charged research, it is not always possible to be a team member or be entirely embedded. This case forced me to question both the limits of my embeddedness in the field as a researcher and the role of my political identity as a Turkish citizen.

Unintended impressions

After a three-hour journey from the Istanbul city centre, I arrived in an area where big skyscrapers and business towers stand next to textile factories in a modest, conservative and slightly poor-looking neighbourhood. The bus ride showed me another face of Istanbul, which I had heard of but had not seen very closely before. The neighbourhood, called Başakşehir, is a newly established part of the city. Previously a village outside of the city, rapid construction had turned it into one of the new faces of the so-called 'change' and 'strong Turkey'.[6] This neighbourhood is also seen as a stronghold of support for the AKP and Erdoğan. In turn, Erdoğan has also declared his support of the neighbourhood on various occasions.

My interview partner, Sermet, was a local to Başakşehir and proud to be a part of this neighbourhood. He stated that he had been able to observe the changes that had taken place there and was also involved in the political youth work of the AKP. He described himself during our talk as a 'strong-willed person who can resist other attractions' – meaning the temptations of a less religious lifestyle, which he had seen during his master's studies abroad, or other political orientations (interview with Sermet, 25 May 2016). During the talk, I realised that he was stressing that he and I come from different backgrounds and that he assessed my political and religious views by my appearance. He offered his opinion in a kind manner: 'I have four sisters, and all of them wear headscarves. I would not like it if one of them decided to unveil. It would indicate that the person had lost her values. But it would be beautiful if you decided to wear a headscarf' (interview with Sermet, 25 May 2016).

His assessment of my appearance was not only about my choice of consciously wearing or not wearing religious clothing, but also about an insignificant (or at least insignificant to me) accessory: my glasses. So I wrote in my field journal: 'He somehow shows me my limits in this research [...] He

6 In 2017, the AKP used the phrase 'Strong Turkey with Yes (Evet ile Güçlü Türkiye)' for the 2017 Turkish constitutional referendum, which transformed the country's political system from parliamentary to presidential. Please see the campaign video of the party: https://www.youtube.com/watch?v=YpVkRgsuvAw.

said that even my glasses give him a clue as to which political view I might support or which part of society I am from'. During our talk, he also said that he would not send me to his other friends in the AKP youth organisations. I stuttered and tried to ask, with a nervous smile, why. He did not give me an answer, and I did not repeat the question. I noted in my journal later: 'His assumptions on my political view might have played a role in this decision' (author's fieldnotes, 25 May 2016).

In this interview, I felt that I was the one who was being observed, interviewed and assessed. In a way, the researcher became the research object. This experience also showed me that such assessments of me are out of my control. Goffman argues that individuals assess others based on their past experiences, and they put on a performance as if they were actors on a stage. They may or may not be aware of their performance, but through such performances, they will be assessed (Goffman 1959). In a similar vein, the researcher also puts on a performance in front of her/his research subjects, and, without noticing, she/he can be assessed as the 'other' or 'untrustworthy', although it was not her/his intention. In my case, Sermet made an assessment on my political view and identity based on my glasses.

Conclusion

In this chapter, I showed several incidents I faced in the field, which indicate different negotiations and how new identities surfaced and were assigned to me in the field: I started to see myself differently, i.e. I became a bi-cultural researcher; my interest in the researched group was interpreted as political belonging; and seemingly insignificant objects such as my glasses hinted at a particular belonging to a group and played a role in accessing or not accessing people. In particular, these experiences showed me that the important feature of these negotiations was for testing the level of trust between the researcher and the interlocutors. Further, I realised that the researcher's identities can be assigned by the interlocutors to the researcher beyond the researcher's intentions.

Overall, these negotiations indicate the complexity of the fieldwork situation. Indeed, such complexity and challenges based on the researcher's identity can force the researchers to state self-critical questions on their practice in the field. Negotiations of identities can create tensions (Lavis 2010) and doubt of the self (Verdery 2018). However, as the reflexivity tradition accepts that researchers are observers in the world and bring their own background in the research, such negotiations should be perceived as important sources of knowledge claims and analysed as natural parts of field research. In that sense, they are not failures in the field, maybe therefore – considering the researcher is equipped with all the necessary research tools – there are just good and bad experiences in the field, which can reveal good ethnography practice.

Finally, I want to make suggestions to overcome the feeling of failure in the field when faced with such negotiations. Firstly, it is important to be aware that possible negotiations can be faced in the field before entering it. Secondly, it is crucial to know that such feelings of fear, failure, and discomfort are part of the field.

References

Abu-Lughod, Lila. 1991. 'Writing against Culture'. In *Recapturing Anthropology: Working in the Present*. Edited by Richard G. Fox, 137–162. Santa Fe, N.M: School of American Research Press: Distributed by the University of Washington Press.

Chatman, Celina M., Jacquelynne S. Eccles, and Oksana Malanchuk. 2005. 'Identity Negotiation in Everyday Settings'. In *Navigating the Future: Social Identity, Coping, and Life Tasks*. Edited by Geraldine Downey, Jacquelynne S. Eccles, and Celina M. Chatman, 116–140. New York, NY: Russell Sage Foundation.

Chege, Njeri. 2015. '"What's in It for Me?": Negotiations of Asymmetries, Concerns and Interests between the Researcher and Research Subjects'. *Ethnography* 16 (4): 463–81.

Clifford, James, and George E. Marcus. 2009. *Writing Culture: The Poetics and Politics of Ethnography*. Edited by Kim Fortuny. Berkeley Los Angeles London: University of California Press.

Coffey, Amanda. 1999. *The Ethnographic Self: Fieldwork and the Representation of Identity*. London; Thousand Oaks, Calif: SAGE Publications.

Geertz, Clifford. 1998. 'Deep Hanging Out', 22 October 1998. https://www.nybooks.com/articles/1998/10/22/deep-hanging-out/.

Goffman, Erving. 1959. *The Presentation of Self in Everyday Life*. Anchor Books. New York, NY: Doubleday.

Gupta, Akhil, and James Ferguson, eds. 1997. *Anthropological Locations: Boundaries and Grounds of a Field Science*. Berkeley: Univ. of Calif. Press.

Hume, Lynne, and Jane Mulcock, eds. 2004. *Anthropologists in the Field: Cases in Participant Observation*. New York: Columbia University Press.

Lavis, Victoria. 2010. 'Multiple Researcher Identities: Highlighting Tensions and Implications for Ethical Practice in Qualitative Interviewing'. *Qualitative Research in Psychology* 7 (4): 316–31.

Lewis, SJ, and AJ Russell. 2011. 'Being Embedded: A Way Forward for Ethnographic Research'. *Ethnography* 12 (3): 398–416.

Ohnuki-Tierney, Emiko. 1984. '"Native" Anthropologists'. *American Ethnologist* 11 (3): 584–86.

Reiter-Theil, Stella. 2004. 'Does Empirical Research Make Bioethics More Relevant? "The Embedded Researcher" as a Methodological Approach'. *Medicine, Health Care and Philosophy* 7 (1): 17–29.

Verdery, Katherine. 2018. *My Life as a Spy: Investigations in a Secret Police File*. Durham: Duke University Press.

3

Attuning to Alterity: From Depression to Fieldwork

JAKUB ZÁHORA

'There's someone in my head but it's not me.'
Pink Floyd

'Them who once lost their mind will never be normal again.'
Vypsana fixa

I got off the bus and after a few moments, I figured out the right direction. It took me about ten minutes to find the address which was stated on yad2, an Israeli version of craigslist, where the owners of a family house were advertising a small studio for rent. I buzzed at the door and was quickly let into the house. I met a young couple who offered me coffee and then showed me the sizeable room upstairs, with a shower, toilet, and a separate entrance from the street. It seemed great. Not only the particular room which would suit me quite well, but also the Israelis who were renting it and seemed fairly nice and communicative, qualities which would be perfect for my fieldwork. Even more importantly, the location fitted the criteria of my research in Israel/Palestine exactly: we were in a small settlement in the West Bank, a sort of community that I set to research for my doctoral project on depoliticisation of contested spaces. However, I did not take the room, nor did I take any of the other three that I went to see over the next few days.

The reason was fairly simple: after seeing each of them, I quickly realised that my mental state would run a significant risk of deterioration. By that point, I had been experiencing depressive episodes for more the ten years. I could immediately foresee that the environment in the settlement, and the loneliness the stay would entail, would likely lead to psychological complications on my part. This decision would, according to many standards, constitute a missed research opportunity, or even a failure – by almost any conceivable academic criteria, I *should have moved* to the settlement. My apartment hunting and the surrender was thus one of the first instances in which my mental state shaped my fieldwork in Israel/Palestine. In what follows I show that attending to my mental condition is necessary to really grasp the contours of my research, how it proceeded, and how I approached it.

Essentially, then, this piece is an attempt to come to terms with what a lot of people would call 'mental illness', namely depression, and how it intersects with my academic work. In writing and publishing this chapter, I hope to show that the personal and the academic are intimately intertwined: my encounters with depression proved deeply formative for my intellectual and later academic self, with tangible repercussions for my fieldwork as well. Through disclosing some of my experiences, I want to show that our research, prisms we adopt, and approaches we take are inseparable from matters often bracketed off in the academic discourse as personal and irrelevant. Contrary to this trend, I illustrate that my mental condition is one of the most important 'facts' impacting my work.

I want to start with a disclaimer of sorts because I am well aware that my issues were never serious to the extent to which they had been for many others. Still, they proved repeatedly disruptive for my life and the lives of people around me, and as such I feel it further justifies my efforts here. However, for me it is also an exercise in coping with these issues. Although they represent only a particular facet of the whole condition, I have gradually realised that thinking about the impact my condition has on my work enables me to objectify my mental states, detach myself from these experiences and make sense of them. Perhaps writing this piece is an attempt at solidifying this analytical distance.[1]

What follows is composed of various segments that I have written over the course of the last four years or so in various states of mind and being, as well as sensations, experiences, and memories that I have not kept in a written form. I've come to think about all this as a peculiar type of archive, an archive that can be organised and from which different files can be extracted so that they can contribute to the narrative I want to offer here. I suppose all life experiences can be understood as such an archive, and indeed this is how social scientists often treat others' lives. But for some reason, I had never

1 Curiously, in this regard the present text and the personal effort to make sense of my condition that animates it somewhat parallels the ethnographic project of learning from a close encounter with 'the field', yet seeking to maintain a distance from it at the same time. I am grateful to Katarina Kušić for pointing this out to me after reading an earlier draft of the chapter.

really thought about mine in this way until my fieldwork. What I realised as I was writing these various segments is that my reflection on my mental health led me to reorganise the way I approach my self.[2]

Illness and me

Spring 2016, Ariel

I first got depression when I was 17 years old. The event (or rather a series of events) that triggered this state of my mental health, this state of *me*, was quite banal as I can clearly see now (or rather as my current self understands it). In fact, it seems so silly that despite my decision to write this piece, I still cannot push myself to talk about it. In any case, at the time, it meant a great deal to me. And as I realise now, it was a transformative experience in many regards.

My condition was not too serious. I was exhibiting some classical symptoms like inability to sleep properly, anxiety, and absence of appetite, but I could still more or less function on an everyday basis. For several months, I was seeing a psychiatrist which I felt was helpful. He provided some food for thought which I seemed to digest quite well, and depression left me in the course of half a year or so. What stayed with me was the memory, a trail, an imprint of the intensely bodily, yet somehow also vaguely mental feeling. I realised later that those would be somatic expressions of my condition, of 'a mood disorder marked especially by sadness, inactivity, difficulty with thinking and concentration, a significant increase or decrease in appetite and time spent sleeping, feelings of dejection and hopelessness, and sometimes suicidal thoughts or an attempt to commit suicide' as defined by the Merriam-Webster dictionary.[3]

These sensations returned repeatedly. Gradually, I discovered that 'normal' pain, i.e. physical pain, could help to alleviate the mental dislocations and anguish. I started to cut myself from time to time, although, in general, the intervals were a year-or-so long. Cuts on hands proved (unsurprisingly) too visible and tended to raise questions from people around me. I thus resorted to cutting my legs – the cuts could then be covered by trousers (I usually cut myself on calves), or I could come up with a reasonably plausible story about them being the result of me running through bushes (I like jogging) rather

---

2 I re-read this passage more than two years after I wrote it. I remember the feeling, almost psychosomatic, that I had when writing it. It was in the middle of my doctoral fieldwork which I was conducting in Israel/Palestine, and I was deeply frustrated with the scholarly literature which seemed so detached from my experience. As I discuss below, I also felt unwell mentally.

3 https://www.merriam-webster.com/dictionary/depression#medicalDictionary

than a product of a chemical disbalance in my brain.

This practice of mine has never been uncontrollable, nothing like the images of bodies covered in self-imposed scars. Mine were very modest scars, almost innocent. Nevertheless, in summer 2014, my condition escalated, and I started to take anti-depressants – I am still using them at the time of writing – as well as attending psychotherapy (although with some breaks). The treatment usually proved to be quite effective, but not always.

Actually, I felt the first signs of this condition as I was walking around Ariel, an Israeli settlement in the northern part of the West Bank, where I was conducting my Ph.D. fieldwork.[4] I felt the things that usually precede and accompany depression, mostly loneliness and self-disgust. As I was passing by fireplaces (it was Lag BaOmer)[5], I started to think about writing down these thoughts and feelings. Obviously, this is a rather established way for some people to deal with these conditions, but I have never tried it. And being trained in political science, I started to link them to what would usually be understood as academic work.

In his work on firemen in the US, Matthew Desmond, drawing on a Bourdieusian conceptual apparatus, suggested that in the course of his participant observation his 'body became a fieldnote' (Desmond 2006, 392). Although Desmond talks about the embodied nature of a particular habitus, I found this remark intriguing with regards to my own experience. I do not mean to say that the somatic experience I went through would provide me with insights into the operation of certain rationalities I was looking into, in a manner parallel to Desmond's research. It is rather that, first, my body became a site for the scars (personal jottings of sorts) which continue to remind me of the very real possibility of slipping into the zone in which a certain amount of physical pain poses as a preferred alternative to the mental anguish.

Autumn 2017, NYC

But Desmond's remarks importantly, and further, speak to impressions which accompany depression for me.[6] I have not written in this document for quite some time. I am sitting in the NYU library. I feel on the verge of depression, like it is within my grasp, or rather the other way around. It is again this

4 This whole section was written in spring 2016.

5 Lag BaOmer is a Jewish holiday which is in Israel traditionally celebrated by lightning bonfires and barbecuing.

6 The following part was written in autumn 2017 when I was a visiting Ph.D. student at New York University.

physical sensation, something that creeps around my lungs. If I am to describe the state in somatic terms, I resort to suffocating. I suppose this is as close as it gets to conveying the physical sensation that accompanies depression. In an article about the experiences of the illness, Andrew Solomon (Solomon 1998) describes his experiences with depression 'as though I were constantly vomiting but had no mouth'.

Autumn 2017, NYC

I thought I would never experience this again. The intensity is interesting. I want to die.

***

In the following months, my condition improved quite significantly but in spring 2018, after I submitted the first draft of my Ph.D. thesis, I slipped into depression again. I was finishing my fellowship at NYU and I decided to take a trip to the West Coast before my US visa expired. Already before leaving, I was not feeling entirely okay but only upon my arrival to Seattle did I realise the severity of the situation. I was unable to focus on anything but the pain. I suppose the new, unfamiliar environment (by then, I felt quite like at home in NYC), coupled with the temporary lifting of the doctoral thesis burden and remembering some taxing personal issues I was going through in the autumn led to a renewal of the depressive state. My original plan was to go to Portland, Oregon and then to hike alone in a nearby national park. At one moment, I suddenly realised that I was quite likely to kill myself if I was to spend several days completely alone in the wilderness. I had never actually felt that I was so close to suicide.

Going back home to Prague a couple weeks later helped significantly but the progress was fragile. In late June, I attended a friend's wedding in the countryside. At one point, I found myself in a kitchen staring at knives and imagining the physical sensation of plunging one of them in my throat and the resulting loss of blood, consciousness, and ultimately, life. Latour once asked 'who, with a knife in her hand, has not wanted at some time to stab someone or something?' (Latour 1999, 177). But for me the question really is who, with a knife in her hand, has not wanted at some time to stab herself?

Ideas like these do something to you. I keep being surprised when people tell me that they have never thought about suicide. For me, these contemplations have become perhaps not everyday, but still consistent parts of my inner life. Despite chemical treatment and psychotherapy, I still find myself thinking from time to time that perhaps killing myself would be preferable to continuing to coexist with these sensations.

These are curious states of being. As I write this, I remember being this person who has these states, these feelings, and these urges. And yet, they also do not really register as my own memories. This is why, on an analytical level, I find these mental states deeply intriguing. It is an opportunity to take a completely different position in and towards the world than is the norm for me. With the risk of exaggerating, I am inclined to say that during these depressive episodes, I do become someone else, someone whom I have a hard time recognising now when I remember these periods.

Illness, academia, fieldwork

With the advantage of hindsight, I realise that the experience of depression had profound influence on my academic work in several distinct regards. This very categorisation, this move of making my feelings neat, is a legacy of my university training: my being honed to identify and pin down social phenomena, label them, and put them into proper boxes. But this reflection does not change the fact that I came to think about these issues in such terms.

First, my alienation from myself was instrumental in drawing my attention to the existence of a multiplicity of incongruous perspectives in the social world we live in. Let's consider what I wrote above: 'I want to die.' Perhaps due to biological imperatives, most human beings do not seek death. On the most basic level, I was forced to embrace that there are indeed people who want to commit suicide, and not in an abstract way: I – or rather a certain form of my self – was one of these people.

Considering that it was me who wrote this and who is yet still alive brings about quite curious ideas regarding different perspectives. I remember having these urges and feeling this pain but at the same time, they are completely alien and incomprehensible to my current self: I cannot relate to the person who had these impressions, although this person was actually me. This bifurcation, the experience of what I now perceive not only as an 'abnormal' urge to kill myself, but also as a completely unrelatable feeling, is quite destabilising. I cannot but acknowledge that I went through this state of mind, yet it is completely alien to me.

The distinctions between my 'current' self and my 'depressive' self, and the repercussions of these experiences, have meanings which go beyond the 'emotional'. During the state of mind that I described above, I adopt a completely different outlook and my ontological and epistemological position shifted profoundly as my being in, and perception of the world underwent quite major transformations. During the worst depression I experienced, time basically stopped; I was sure nothing was to ever change; I could not imagine the pain would go away; and I could not focus on anything but it. I suppose this is not dissimilar from what Elaine Scary (Scary 1987) famously called 'world-destructing pain', although inflicted in a different way. I essentially went through a dissociation from my former self not only emotionally but also in terms of how I understand the qualities of this self and the environment I find myself in. What this means is that I somehow had to come to terms with the fact that there are things which are just on completely different planes of reality, impressions and experiences which are not reconcilable, and yet which coexist not only within the same physical space but even within the same person.

As a result, I believe my experience with depression led me to embrace particular academic approaches known under the label of 'poststructuralism', or at least it significantly facilitated this process. Although this claim might be a bit contentious, I would say that one of the main premises of what became known as critical scholarship is to recognise that there is a multiplicity of prisms, world views *and* ways of being in the world, and by extension that one's position is necessarily just one contingency among many other possibilities. To really accept this is, I would say, actually a fairly hard task – it entails realising the confines of one's positionality and its arbitrariness, something that is antithetical to the established notion of the self. Indeed, achieving this realisation is something that we seek to achieve through the laboured process of fieldwork that is supposed to bring us closer to the worlds of others.

I feel that depression hugely aided this process in my case. I do not doubt that I could have come to this experience through different means, as many others have. Nonetheless, it was the personal, even intimate experience with depression which made me attuned to alterity. This bifurcation – emotional, psychosomatic, epistemological – brought about by my mental illness made me aware of difference and otherness in a way that solely intellectual journey could not. As I am writing this, I cannot really relate to my own self from a few years earlier. What I am trying to say is not only that this realisation casts doubts on my ability to understand others (along the lines of 'how could I if I can't even make sense of my own thoughts') but rather that it made me more open to the multiplicity of life experiences in the first place. I did not only comprehend that different people have different outlooks and worldviews; I lived this difference.

In a distinct sense, I think that my depression had a more concrete and specific impact on my scholarly conduct which I touched upon in the opening lines of this chapter. This transpired mostly during my doctoral studies, especially during my Ph.D. fieldwork. Not unimportantly, my mental condition made my stay in Israel/Palestine even more taxing. From several conversations I had with other researchers, activists and foreigners working in the region, I would argue that it is safe to say that the local situation is not exactly beneficial for one's mental health. I guess it is redundant to say that already having issues in that regard does not help.

As a result, my mental illness deeply impacted the way in which I conducted my fieldwork. There were several concrete instances in which my condition made me unable to take steps which I knew would be beneficial for my research but fairly detrimental for my well-being. I opened this chapter with one of these cases – I could not imagine living alone in a small, apparently boring, community without much social contact, even if it could have been useful for my doctoral project. Similar considerations also shaped my decisions several months later, after I spent some time in Ariel, a bigger settlement. My original plan was to spend the rest of my stay (about four more months) in Israel/Palestine in the settlement in order to further my 'first-hand experience of everyday life' in this particular field. However, I was struggling not only with alienation from my informants but also with reoccurring periods of mental instability – part of this chapter was written during that period as an attempt to offset these crises. I knew that I should stay in a settlement for academic reasons. But I also knew that I would be fairly miserable because of that. In retrospect, I am glad I chose to leave, even if I missed more opportunities for immersion.

Later I noted in my thesis that I simply 'didn't have what it takes' (D'Aoust 2013) to fully immerse myself in the community I came to study. I still think this was the case. But I also realised there are limits to the lengths to which one should go in order to 'have what it takes', to be a 'proper' fieldworker, and to do the kind of research which would expose oneself to harm, physical or psychological. Based on my experience, I believe that these considerations should also matter when one tries to figure out the parameters of her stay in the field.

Later, my mental issues proved highly harmful for my academic productivity, something that is rather unsurprising to anyone familiar with these conditions. During the last year of my Ph.D. when I was a visiting student in New York City, my condition worsened for a few months, to a fairly debilitating extent. I could still function in terms of the everyday, less demanding activities, but I was not able to write. I did not think I would be able to finish my thesis if the situation persisted.

Nonetheless, in hindsight I can also see that my condition had imports which facilitated my research. In the course of my stay in Ariel, I was having a hard time really following the core ethnographic commitment and understanding the world of people I came to work with. I felt alienated from Israeli settlers with their conservative, right-wing, and casually racist remarks. Even more so, I could not really comprehend their mostly indifferent attitude towards the larger political conditions they were part of. In short, this constituted what I perceived as an ethnographic failure, a failure to really relate to the research participants.

It was only after I finished my fieldwork that I managed to harness empathy for the Israelis who move to the West Bank to live among the occupied Palestinians without giving much thought to the system they effectively maintain. First, I became wary of my own position as a foreigner who came to Israel/Palestine to get enough 'data' for a project, and then left once this task was completed. As such, the extractive nature of my stay on the one hand, and taking advantage of socioeconomic benefits by the Israeli settlers on the other, seem not so different. Also, as I discuss elsewhere, my attention to various material and visual practices that condition the everyday life in the settlements helped me to better come to terms with the epistemological, political, and ultimately ethical gap between the settlers' experiences and the nature of Israeli occupation.

But I would also say that the attitude shift vis-à-vis the settlers on my part was enabled by my previous close exposure to Otherness – not political difference but the alterity that occurred between different states of me. It was, in a way, perhaps the ultimate ethnographic experience, an experience that made me more receptive towards other ways of thinking, feeling and being, albeit only after I finished my fieldwork and could better reflect on the encounters I had in the settlements. I don't think I could understand the settlers to a similar extent, seemingly so different and detached from me, had I not encountered difference and detachment from and within myself before.

Illness and failures

There is an infamous tendency to treat mental illnesses as individual insufficiencies, as particular *failures* of will. This is naturally not the kind of failure I want to entertain here. The existing research clearly shows that one cannot conceive mental health issues as personal shortcomings, something that I suspect the readers of this particular volume are aware of. And there is now more than enough evidence that higher education is a particularly hostile environment for people with mental illnesses – and that academia even induces those.[7]

The failure that I want to briefly talk about here is essentially two-fold, and to some extent consists more of repeating what others have expressed much better before me (see e.g. Hartberg 2019). First is the absence of institutional support for people in academia struggling with various mental illnesses (which reflects the situation in our society at large). Although there are some consultation services available, they are usually insufficient to really provide support in the midst of the current mental health epidemics. Frequently, these concerns are even more pressing during fieldwork when the researcher finds herself away from what (hopefully) had become an academic home where one can find refuge among colleagues and friends. In the field, the pressure of research is thus combined with the absence of social and institutional safety nets.

Second, failure is a somewhat more specific articulation of the first one. It is, very much in the spirit of this volume, failure to publicly discuss how these and similar experiences and issues impact our scholarly conduct. By now, it has (thankfully) become standard in the academic works based on fieldwork to attend to one's positionality and how it plays out in the field. Nonetheless, to my knowledge there is a lack of works which would attend to mental health in relation to fieldwork (for an exception see Tucker and Horton 2019). As I discussed in this chapter, these issues are crucial, not only in terms of individual wellbeing but also in making our knowledge claims. If we accept that the research experience is an embodied one, we need to attend to these aspects as well: for many, mental issues are part and parcel of the stay in the field.

Conclusion?

After Katarina read the first draft of this chapter, she asked me what I was trying to achieve by writing all of this, 'what am *I* hoping to get from publishing the chapter'. When I was thinking about this question, I realised that the most honest question is rather selfish. As I wrote in the introduction, I sought to adopt detachment from my condition: over the last few years, I found out that writing down my impressions and taking a certain distance, turning my depression into (yet another) problem to be probed academically, helps me to

---

[7] For example, a recent study (Levecque et al. 2017) found that 32% of Ph.D. students are at risk of developing a mental disorder, mostly depression, a rate much higher than other comparison groups. Importantly, this applies not only to early career researchers, but senior staff (see e.g. Weale 2019) as well as non-research students. With regards to the latter, according to research conducted among the UK students by *All Party Parliamentary Group on Students* (All Party Parliamentary Group on Students 2017), one third of the respondents reported having suicidal thoughts at some point.

better cope with it. Doing so does not allow me to overcome the condition altogether but it is useful for making sense of this experience. It felt that publishing what I wrote over the last several years is a logical next step, a step which would further cement this attitude.

But beyond this, in this chapter I sought to caution against disregarding the importance and salience of 'personal' and health issues for the academic conduct, practice and knowledge production. Not only do my experiences show that the academic is intimately related to other dimensions of one's life, they also demonstrate that we need to weigh academic calculations with concerns for our wellbeing, perhaps especially in the context of a taxing stay in a foreign and strange land of 'fieldwork'. And I also hope that in discussing these issues, this chapter can be useful for people facing similar problems, although the pessimistic me suggests that it might be wishful thinking.

I am sceptical towards the promise of this text because I recognise that it is easy for me to say these things now, when I have not experienced a depressive episode in a while. The following (and concluding) paragraphs that I wrote more than a year ago show that my current stable state is far from guaranteed, and that its deterioration can have a serious impact.

New York City, May 2018

Andrew Solomon (1998) finishes his piece on coming to terms with depression by noting that 'I cannot find it in me to regret entirely the course my life took'. As I read it, I remembered that I uttered almost the exact same words some 13 years ago, in conversation with a friend after I had undergone my first wave of estrangement from myself. Perhaps this piece is a reiteration of this sentiment: what I wanted to show here is that my current self and my depression cannot really be separated, and in many regards I am glad for the forms that my self and, with Solomon, my life at large obtained.

But at the same time, I am not sure if the price was not too high. My depression was (or rather has been, and perhaps will be again), after all, extremely painful. And they proved to be disruptive for the lives of people around me, in some cases they served as a large part of the reason why people I considered close became estranged from me. This is why I feel that writing all of this is also, or perhaps mostly, an effort to be able to cling to this analytical self the next time I feel my sanity waning and dissolving. But I am afraid it might not work. Because with every new seizure, I feel my resolve to struggle against these disruptions fading a little bit more. I am just tired. I will stop now.

**The author would like to thank Katarina Kušić and Františka Schormová for their comments on previous versions of the chapter. All that you dislike about this text is my fault though.*

References

All Party Parliamentary Group on Students. 2017. 'APPG Briefing: "Lost in Transition? – Provision of Mental Health Support for 16–21 Year Olds Moving to Further and Higher Education."'. http://appg-students.org.uk/wp-content/uploads/2017/07/APPG-on-Students-December-Mental-health-briefing.pdf.

D'Aoust, Anne-Marie. 2013. 'Do You Have What It Takes? Accounting for Emotional and Material Capacities.' In *Research Methods in Critical Security Studies: An Introduction*, edited by Mark B. Salter and Can E. Mutlu, 1 edition, 33–36. London and New York: Routledge.

Desmond, Matthew. 2006. 'Becoming a Firefighter'. *Ethnography* 7 (4): 387–421.

Hartberg, Yasha. 2019. 'Depression and Anxiety Threatened to Kill My Career. So I Came Clean about It'. *The Guardian*, September 10, 2019, sec. Society. https://www.theguardian.com/society/2019/sep/10/depression-and-anxiety-threatened-to-kill-my-career-so-i-came-clean-about-it.

Latour, Bruno. 1999. *Pandora's Hope: Essays on the Reality of Science Studies*. Cambridge: Harvard University Press.

Levecque, Katia, Frederik Anseel, Alain De Beuckelaer, Johan Van der Heyden, and Lydia Gisle. 2017. 'Work Organization and Mental Health Problems in Ph.D. Students'. *Research Policy* 46 (4): 868–79.

Scary, Elaine. 1987. *The Body in Pain: The Making and Unmaking of the World*. Oxford: Oxford University Press.

Solomon, Andrew. 1998. 'Anatomy of Melancholy'. *The New Yorker*, January 4, 1998. https://www.newyorker.com/magazine/1998/01/12/anatomy-of-melancholy.

Tucker, Faith, and John Horton. 2019. '"The Show Must Go on!" Fieldwork, Mental Health and Wellbeing in Geography, Earth and Environmental Sciences'. *Area* 51 (1): 84–93.

Weale, Sally. 2019. 'Higher Education Staff Suffer 'epidemic' of Poor Mental Health'. *The Guardian*, May 22, 2019. https://www.theguardian.com/education/2019/may/23/higher-education-staff-suffer-epidemic-of-poor-mental-health.

Part II

Situating Knowledge

4

Failing Better Together? A Stylised Conversation

JOHANNES GUNESCH AND AMINA NOLTE

Failure, naturalised and de-constructed

Amina: How paradoxical it feels to entertain the notion of failure while we are sitting under the sun, next to a pool in which a plastic crocodile is floating.

Johannes: Yes, paradoxical, but maybe also quite telling.

A: Why?

J: Because for early-career scholars like us, failure might reveal itself precisely in the fleeting luxury we enjoy sitting next to a pool.

A: Haha, true, aspiring yet full of insecurity.

J: So, the pool is actually a good place to start a conversation about what we might mean when we speak about failure.

A: And how we might fail better together!

J: Yes, but how do we do that? Where would you start?

A: I have this impulse to de-construct failure. As it seems to be all around us, this could help us question the centrality failure has assumed in academia.

J: Interesting, my first impulse would be to conceptualise failure. This, at least, is what I am trained to do. In political science, we often proceed deductively. First, we name the beast, then we try to tame it. So, what is failure, which failure are we talking about, where do we locate it? And then: What do we do about it?

A: Funny that you give so much credit to your disciplinary background. In anthropology, you hang out first and see where it takes you, without having to determine everything beforehand. Is this not a bit of a contradiction?

J: Yes, maybe, but it is actually my ethnographic work that triggers those abstract thoughts.

A: How so?

J: I am currently trying to make sense of my empirical material and question how and if at all I can bring different facets together. In reflecting a bit on the practice of ethnographic work in political science and international relations, I realised a tension that I find intriguing. This tension is particularly pronounced in critical research, I think, especially if critical means to question what is, how it came to be, and could be different (see Sjoberg 2018).

One way to illustrate this tension, and look at what it does to ethnographic work, is to probe naturalisation and de-construction (cf. Webster 1986; van Wingerden 2017). By naturalisation, I mean the act of attesting, describing and thereby determining 'what is'. It is a necessary and positive requirement of all communicative practices, including research interactions. For example, now we talk about 'failure', what it means to us, and thereby circumscribe its 'nature' (good or bad, productive or destructive and so on). Likewise, in fieldwork-practice, we try to make sense of other people's sense-making and capture phenomena 'on the ground', wherever this might be. We therefore inevitably naturalise when we generalise from particular experiences. And this is precisely where de-construction comes into play, which seeks to strip things of their assumed naturalness and to probe contingency, diversity, and emergence.

Now, if fieldwork is to be critical, distinct requirements overlap: The 'fieldwork encounter' valorises 'being there' (Borneman and Hammoudi 2017), while the 'spirit of enlightenment' aspires to see beyond and overcome our 'self-inflicted immaturity'. The problem is that those two requirements are simultaneously mobilised to legitimise research, assume authority, and thereby determine success and failure.

A: Ok, so the tension you describe is productive in that it shapes the practice of ethnographic research?

J: Exactly, and this is why such a seemingly dry methodological matter is actually deeply political (see Marchart 2007). Lest we forget, there is not one final cause, God is dead, reason sometimes wicked, utility not monolithic, capitalist growth endless, and so on. This is de-construction, if you will, which does away with clear-cut criteria for failure, too. Nonetheless, in pursuit of empirical validation, professional recognition or confirmation of expertise, we all partake in the act of foundation. We try to establish authority. Those acts are plural and provisional, but they render certain meanings and social artefacts more efficacious than others: Linking productivity to profitability is particularly conspicuous here. After all, progress is paramount and 'success needs to be earned', which generates competition. Thus, critical research simultaneously scrutinises and manifests differences; and any attempt to resolve those differences, for example through disciplinary fiat, animates the ensuing contestation.

A: Hmm, you said it yourself, but this really sounds very abstract. What does it bring to our account of 'failure' in fieldwork and critical research?

J: Two things: That 'failure' cannot be resolved, if only because it means very different things for different people. And that this needs to be worked through. To give a paradigmatic example drawn from the 'correspondence' and 'consensus' theories of truth, respectively (Jackson 2010): For a positivist perspective of mind-world dualism, failure depends on whether or not research 'corresponds' to the 'real world'; for a post-positivist perspective of mind-world monism, it depends on resonance with interlocutors. The former is 'objectively' determined by the use of statistics for example, the latter contingent on 'inter-subjective' understanding.

This also means each perspective has different requirements of naturalisation and de-construction. Yet, despite those differences, we are all trained and expected to 'get it right', no matter our disciplinary affiliations. This makes 'failure' ominous and 'success' a persistent expectation.

Failure or not?

A: But you seem to accept 'failure' as a term.

J: You would go further?

A: Yes, I would question why we are giving so much weight to failure, especially in the context of ethnographic fieldwork.

J: Why do you think this is? Why do we approach fieldwork through the lens of failure?

A: I think it is partly because, as young researchers, it has become so much part of our experience, our daily environment, and our thinking. The world we live in – the precarious academic world we move in – produces us as failing subjects. Upon our initiation, we are introduced into a world in which we already fail. As a consequence, ambitious as we are, we have learned to accept failure as a term, as a concept, as a state of being. It looms in the background; a price we feel we have to pay for doing 'what we love'.

J: This is our metaphoric crocodile.

A: Yes, the crocodile of academia if you want, where failure looms in the background of everything we do: of every application and proposal we write or end up not writing; of every interview we are being invited to or not; of every beginning even – be it a sentence, a paragraph, or an entire chapter. Failure is not being able to pursue, to produce, and to perform.

J: So, failure is a necessary component of neoliberal academia?

A: Yes, of course. We are taught to anticipate and manage 'failure'. Ultimately everything is 'trial and error', working to 'fail better next time', as Beckett said. Towards that end, failure and the anxiety to fail have become our companions, so much so that we can only learn to handle our precariousness as 'adeptly as possible' (Lorey 2015). But this inadvertently makes us lonely. To struggle with vulnerability, we focus on ourselves. We tend to apply, write, submit and publish alone to distinguish ourselves. We thus also risk failing alone, which is why we have to work harder, alone. To become more mindful. More aware. Resilient. We turn to 'self-help'. So that failure makes us stronger. And, in the end, failure becomes the premise on which (academic) success is supposedly built.

J: But what does it mean to start something on the premises of failure? Can we reject failure as a term, as a concept? Doesn't failure imply knowing what non-failing would entail? What is in between failing and succeeding? An experience, a conversation?

A: Exactly, and this is what ethnographic fieldwork is all about: experiences, engagement, and exchanges – and, most importantly, the reflection on those experiences. Thus, to me it is a question of how seriously we take ethnography in all its dimensions: as a practice of inquiry into social worlds of which we as researchers are an inherent part; as 'actively situated *between* powerful systems of meaning' (Clifford 2009, 2) in 'which human ways of life increasingly influence, dominate, parody, translate, and subvert one another' (Ibid., 22); as a process and an open engagement with the simultaneity, multiplicity, and ambiguity of lifeworlds. If we take all those things seriously as we claim to do, why do we relate fieldwork to the boundedness and fixedness of failure?

Failure is ubiquitous

J: Because failure is annoyingly ubiquitous. In my own research, I am continuously confronted with failure. It is there, we cannot just think it away – even though I would be very sympathetic to this kind of undertaking.

A: Could you give an example?

J: Yes, several. For one, there is failure because human suffering is a reality; inequality, poverty, and violence are real, and they are aggravated by discursive constructions that decide whose voices are heard and whose suffering is recognised. This is a failure of politics and proof of our complicity in it. As such, I constantly fail because I do not want to hide behind some form of moral relativism. In my research, I trace the resonance of the Egyptian uprising in international development cooperation, where I also worked for some years. In particular, I focus on how the basic demands for 'bread, freedom, and social justice' are negotiated, mis-appropriated and thereby *dis*-qualified. As all those demands are put forth against the reality of economic marginalisation and political disenfranchisement, the problem is that my research entails a double blind: not only do actors in development cooperation tend to disregard people 'on the street'; in scrutinising what those powerful actors do, I also confirm their prerogatives and the exclusions this generates. Thus, I wonder to what extent I actually contribute to the cause of 'bread, freedom, and social justice' through my research – or whether I am not also undermining it, no matter how critical my undertaking claims to be.

Second, not only do I fail to live up to my ideals, but arguably also profit from the ensuing situation, which is very unsettling. As a white, male academic, I got a position at the UNDP without any particular knowledge about Egypt when I first arrived. Then the uprising happened, and I got drawn in until this day. Yes, I learned a lot, about political mobilisation and organisation, the politics of international solidarity, about myself… But if it wasn't for the 'failure' of the Egyptian uprising and the misery it has brought upon so many people, my research would probably be only half as appealing. And now I can 'use' my experiences as a commodity in the academic market, not only to sanction my conclusions, but also as a competitive advantage. After all, I have *valuable* first-hand experiences with the uprising and development cooperation. I am being cynical here, but I profit from the Egyptian uprising in ways that most Egyptians do not because the entrenched structures of capitalist exploitation work in my favour.

Third, there is failure because the confrontation with authoritarianism has wide-ranging repercussions, also in research practices. In and beyond Egypt, insecurity, fear, mistrust, anxiety, and violence are widespread. This affects who you talk to, how, what information people relate… Secrecy and gatekeeping are common, and not only because of malicious intentions. In the most extreme cases, people simply disappear, are put in prison or murdered. This happens to Egyptians by the thousands, but nowadays also to foreigners. Giulio Regeni's tragic death is demonstrative (Nassif 2017; Palazzi/Pusterla 2018). The ensuing outcry over the murder of a foreign researcher inadvertently exposed not only the precarious and perilous politics of knowledge-production, but also its entanglement with the political economy of authoritarianism. That is, the structures of capitalist exploitation are related to hierarchies of signification. They affect whose lives, ambitions and sufferings are recognised or not (cf. Butler 2006). And they circumscribe to which people, experiences and narratives we have access to. This depends on the requirements of ethical research practice, but also racist, sexist, classist divisions. In my case, I had to cancel my fieldwork in Egypt. Instead, I now seek to trace the resonance of the Egyptian uprising beyond Egypt. By focusing on development professionals and fellow researchers, I try to put the critical gaze on those with privileges that many Egyptian activists don't have (anymore). But at what costs?

Fourth, I know failure because it underpins every step I take. Here, I fully agree with what you said before. We live in an environment that generates competition, induces precarity, valorises commodification, individualises responsibility, and thereby raises us as failing subjects. I actually find academia particularly odd: While there is an abundance of critical engagement with neoliberalism, young researchers oftentimes comply with the basic neoliberal requirements: We are ever-mobile and risk social relationships while we are at it, work non-stop for meagre pay and petty benefits, don't unionise, let alone properly mobilise ... and blame ourselves if we don't make it after all. In all earnest, I heard people tell one another to 'suck it up', 'toughen up', 'it's part of the game' and so on.

Ok, let's rephrase failure

A: Amen, but I would not call any of those points failure.

J: What would you call it?

A: I would first want to ask: what leads you to think about them in terms of failure?

J: Hm, I think it takes a lot not to internalise the regimen of failure when it is constantly rubbed into your face, to say the least.

A: Yes, but without falling into the trap of 'positive psychology', I think that what you mention above is a very productive awareness of the pitfalls and possible dangers of ethnographic fieldwork; of the delicate and sensitive situations that we engage with as researchers; of the harm we often cause without knowing or through the will to know. But reflecting on all of the above is not failure. It is taking seriously ethnographic practice, political context, and the situatedness of experience and knowledge; it is recognising the importance of solidarity, but also the limits of representation. As I said earlier, I think we miss the complexity of working ethnographically in the field if we approach it in terms of a binary distinction between success and failure, right and wrong, complete and incomplete.

J: So how did you experience that during your own fieldwork on urban infrastructure and its contestation in Jerusalem?

A: I had to learn it the hard way. I started my fieldwork by being scared to fail, but I ended up abandoning the term from my own research vocabulary. Now, I think one cannot fail in ethnographic fieldwork. I came to think of it as an open process from which I take what I am able to observe and reflect on it. I mean, there is so much happening 'out there', all at once. But our ability to see and not see things, and to work through them, is limited by how we learned to see and unsee. To accept the partiality and limitedness of one's own perspective is a big chance and relief – but to some it might appear as failure, I guess.

J: How did this realisation come about?

A: Initially, I had no idea what this fieldwork would look like, where it would lead me, what I expected to take away from it. But when I moved to Jerusalem, I realised that there was no field – and no clear-cut failure or success either. I did not enter a 'field' when I entered Israel. There was no 'beginning' of fieldwork and no 'ending'. All I found was a shaky continuity: a continuity of an experience, of a journey, and of a conversation.

Instead of thinking of my research as work in a discrete field, I started thinking of 'spacework'. Not because my experience became something out of space but rather in terms of the spatial continuities, frictions, and struggles I learned about (Tsing 2005). I entered that 'space' way before I had physically entered Jerusalem. The boundedness of my research subject dissolved in front of me once I realised that the very space(s) I wanted to research were the ones that I already moved in, that formed my experience and shaped my perspective.

J: Could you give an example?

A: Entering Israel through Ben Gurion Airport always marks a crucial point in my journey through this space. I entered and always enter as the privileged white academic that I am. With a German passport. And a Muslim name – Amina, the mother of Prophet Mohammed. The 'trustworthy', as it translates. But my name is not trustworthy in the heavily securitised space that I enter. The person that I am is not to be believed. I end up sitting in the immigration area in which subjects get securitised through routinised practices of knowing the 'enemy'. Questions. The name of my grandparents, my parents, my siblings. Waiting. Hours of suspension, every time. In silent company with many others, uneasily sharing a space of uncertainty, of subjectification and in between-ness. Not yet in the country – but already in it enough to be subjected to its rules. In it enough to know that compliance helps. Patience. A smile. Some Hebrew words.

J: I can imagine that this experience also shapes every interaction you have during your 'spacework'.

A: Yes, for example every email I sent out as a request for an interview.

Should I change my name? Would my name and my interest in Israeli security practices be too suspicious? Would it change how people approach me? I recall how an Israeli security advisor told me straight away that he had 'checked' up on me before our meeting; how the police commander, responsible for the security of infrastructure in Jerusalem, seemed really alert when I called him to ask whether we could meet; how the actors I tried to follow were all of a sudden following me. Googling me. Reading articles I had published.

This made me think a lot about how affected my 'results' would be from all the presumptions and considerations that the people I interviewed had already gathered about me. How the knowledge they shared with me would already be filtered and weighed. And how I, as the young female researcher, had to comply with their rules of the game. I remember how I played extra naïve during our conversations, not allowing myself to show any disapproval of their words. I remember ignoring the masculinity displayed while talking about security trainings, drone operations, surveillance, and targeted killings along infrastructure in Jerusalem.

J: What did you take out of these experiences?

A: Many questions and maybe some preliminary answers… So, after all, did I fail? I don't think so. Rather, I adapted, but in a political manner. I came to reflect on the hegemonic discourse I settled in, on how the spaces I lived and moved in were permeated by fixed articulations, sedimented by daily practices and routines. This relates to what you said before about naturalisation and de-construction. Seen from this perspective, every coffee, every walk to the grocery store and every movement happened in the growing awareness of how hegemony works: how hegemony produces its own subjects, how it impacts the things we find and do not find, things we hear and not hear, things we see and do not see.

Is this failure? I doubt it. It helped me to refine my theoretical and conceptual reflections, to deepen my initial flirt with Gramsci, Laclau and Mouffe. I understood the value of hegemony as a concept only against the backdrop of my ethnographic engagement. I learned that 'hegemony is never complete' (Crehan 2018, 136) but is always at work through the 'contradictions between the official narratives of the dominant and the actual experience of subaltern' (Ibid.). Hegemony, as I sort of knew before but only really came to understand through the engagement with my ethnographic encounters and materials, works through the everydayness, the mundane, and the common sense.

J: Well put, and very relevant for how failure as a token of capitalism becomes hegemonic, too.

A: Yes, but the point is that it took all my ethnographic work to realise just that. I spent six months in Jerusalem, between Israel and Palestine, in order to research Israeli security practices around so-called 'critical infrastructure'. I was interested in how infrastructure is constructed as critical and what is implied in this construction. How does 'critical infrastructure' affect the people who use it? How do the actors around it understand what they do? How are politics done under the premises of security concerns and practices? How is an entire world and a society built on the vague meaning of security?

It took me a while to realise that everything I did was a part of what I wanted to research. That the continuity between security practices 'here' and 'there' connects spaces and disconnects others; that security is not to be found in a bounded field but rather in and through the spaces through which its diverging meanings and practices move; that security circulates; and that it materialises spatially, fragments spaces of solidarity, and uproots feelings of safety and community.

J: So, with regards to fieldwork and failure, what does that mean?

A: Well, what I am saying is that there is no field, not at least in any clearly bounded way. And hence there is no definite failure either. The idea of a bounded space or a fixed temporal sequence works with ideas of a 'beginning' and an 'end', with clear ideas of who has to reach what, in a specific time and place. Instead, the many things that emerge across and in between the social relations that make up our fieldwork practices matter much more to me.

This brings me back to what I said before, that trying to put up with failure makes us lonely. In my time in Jerusalem and beyond, I have had various encounters with fellow researchers, working on similar subjects and going about their own ethnographic endeavours. However, sharing a research subject only seldomly evoked joy or sympathy in these encounters. Rather the opposite: I realised that the fear of failing is even bigger if there is already someone out there who might have better access, more contacts, more experience, or more publications.

J: Here again, ethnographic fieldwork is charged with anxiety, but different from what you said before.

A: Yes, this is the very particular anxiety of meeting someone who has been quicker, who has been 'there' before you, who harnessed all the information 'out there' and who might be faster in 'using' the information in order to advance their academic career. Here, the 'field' is given a very specific temporal and spatial delineation that it naturally does not possess, and which is at odds with what I said before about it.

The tricky part is this: in theory, we cannot be precarious alone. Being precarious does not exist in itself; it is always relational and 'therefore shared *with* other precarious lives' (Lorey 2015, 12). But in reality, the complete opposite tends to happen. Instead of acknowledging a shared experience and appreciating that our subjective understanding of things will always lead us to see, reflect and write differently than others, we feel endangered by others.

If we were to understand that we cannot fail with what we do and that our research will always be framed through the uniqueness of our own perspective, would this not make us more open to relate to each other? Instead of uniting us then, precariousness separates us. Turns us into anxious individuals. And hence it makes us governable in the sense that we compete with each other about who exploits him/herself most 'productively': for funding, for positions, ideas, publications.

J: What do you make out of this?

A: I think that it obstructs the very openness, sharedness, and resonance that ethnographic fieldwork requires. Anxiety leads to everything that ethnographically-informed research should reject – it encourages gatekeeping instead of cooperation, disclosure instead of open exchange, and silence where there should be flows of words, discussions, phrasing and rephrasing, thinking together and with one another. Thus, the fear of failure makes us fail even harder. Instead of accepting the inter-subjective, the personal and positioned relations that exist between the researcher and their 'research subjects', making every research unique in its own way, researchers compare themselves with each other. Instead of relating to each other, learning from each other, they compete. And this, inevitably, only leads to more failure.

Five inconclusive suggestions for failing better together

J: So, what do we do?

A: Cheeky, that's what I wanted to ask you!

J: I think you are right to point out that we will inevitably fail as soon as we accept the premises of failure.

A: But I also see how anxiety and fear are mechanisms of producing neoliberal subjectivities.

J: Yes, this is something structural, which might not leave us with a lot of possibilities to personalise those pressures. And what is more: if we were to turn those pressures into something positive, we would put it upon ourselves again to adapt, which is precisely how the whole thing works in the first place.

A: But should this stop us from trying harder, or differently, as long as we can at least? Instead of obsessing about our victimhood and helplessness, I really think we need to move on.

J: Where to?

A: Away from our disciplinary routines and the comfort they provide or seem to promise maybe.

J: Ok, let's think, maybe we can identify some very inconclusive suggestions for failing better together.

A: First of all, we could share more. This could help against the commodification of research, the hierarchical politics of expertise, and the uneasy attempt to position oneself as an expert. We can share stories and experiences – positive and negative ones alike. We can share materials, readings-lists, and annotated bibliographies with colleagues. We can co-generate research with interlocutors – instead of informants that contribute 'data', the people we engage with can partake in the conception of research, its writing, and dissemination. But all this requires that we actually try to engage with one another, in conversations, seminars, and supervisions.

J: Yeah, we tend to forget that the struggle can also be beautiful, when we find some shared meaning, a purpose even or a cause.

A: In any case, value is more than just a product – and the purpose of research is not only to come up with a definite conclusion. If we were to talk not only of successes, failure would then become less menacing. And as an incentive to reflect, learn from and gain a sense of purpose in contrast to what we do not want, failure could even be worth experiencing.

J: Relatedly, we could acknowledge the numerous factors that induce vulnerability, which is very different from self-pity. This could be the second point. Instead of artificially separating emotions from research, values from facts, and mind from world, this could help to expose research as personal. Against the ironclad positivist trinity of objectivism, empiricism, and naturalism, there is much to be explored with regards to what our research does to us and our interlocutors. This is what Elizabeth Dauphinée (2010) suggests when her main protagonist asks: 'What expert am I?' To me, this daunting question should neither prompt self-indulgence nor an automatic vindication if only one exposes their tribulations. It would not only be insufficient, but also counterproductive to consider the question in isolation from the socio-political context. Rather, careful scrutiny of structures of domination and hierarchies of signification is needed because they implicate us all. For example, in the course of the Egyptian uprising, fear has long crossed the Mediterranean (Wahba 2018). Now, in many places, politics works through fear, but this is why it could be an unexpectedly empowering act to attest its pervasiveness (Kanafani and Sawaf 2017).

A: Yes. But there is something that makes me feel uneasy with what you are saying. I think it has to do with the emphasis on responsibility or rather the underlying expectation to assume it. Responsibility for what? Where to start? Am I not overly responsible already? Why me? And what about collective responsibility? After all, 'we' care more for those that are close to us, which renders empathy prone to racism. It seems like responsibility has become a disciplinary force, a perversion of Foucault's 'care for self and others'. In this way, responsibility has debilitating and individualising tendencies.

J: True, but maybe this is not so much a problem of responsibility, but its de-political rendition. And here is our third point, I think. As part of it, we need to more confidently establish and clearly communicate standards by which we judge, assess, and act – in ethnographic research as well as other forms of political practice (Schatzki 2009). This goes against my own cynicism as well as (post-modern) relativism. Standards of positivism and policy-relevance are as straightforward as they are predominant, but commitments to justice could oftentimes be made more explicit. And we should utilise our critical purchase. When we feel we do actually get it right, we should speak out, as clearly as possible, and heed the consequences of our insights. For too long violence has been sanctioned and normalised while its profiteers become apologists. But I feel that what is happening around us is too important to be left to those that are audacious or ruthless enough to speak the loudest. Thus, for me, to cultivate responsibility means dealing with the politics of expertise and the uses and abuses of authority.

A: I like this commitment for, rather than against something. This might be the fourth point, namely that it is not enough to complain about failure and criticise neoliberalism, but that we need to do something about it. In academia, critique is performed extensively. Yet, for all sorts of reasons, critique has run out of steam, as Bruno Latour (2004) famously put it. The performance of critique not only remains inconsequential, but it also helps maintain the status quo (Boltanski/Chiapello 2007). As a case in point, critical researchers also play along even though they obsess over neoliberalism and its effects. But austerity and authoritarianism affect us all, from CEU, which is forced out of Hungary, to many other institutions. So how can a greater concern with social justice be purposefully integrated into academia? Maybe we really need to do things differently, build different, better partnerships. But how can those that have the means support those that don't?

J: That's tricky, but I think we need to shift the focus in order to figure it out. This could be the fifth point. There is much more to critical research than publish or perish, success and failure. Supervision, collegiality, and care are crucial. Lively exchanges, feedback, and revisions matter way beyond a footnote. So, thank you, Katarina and Jakub, as well as all the participants of our workshop, for your comments and suggestions to this conversation! And thank you for being open-access, E-International Relations. We need more spaces to cherish the many experiences, curiosities and contradictions that lie between success and failure in research.

A: Yes, and we need to maintain this conversation and think together – with or without a plastic crocodile in a pool, within and beyond academia.

References

Boltanski, Luc, and Ève Chiapello. 2007. *The New Spirit of Capitalism*. London: Verso.

Borneman, John, and Abdellah Hammoudi, eds. 2017. *Being There: The Fieldwork Encounter and the Making of Truth*. Berkeley and Los Angeles: University of California Press.

Butler, Judith. 2006. *Precarious Life: The Powers of Mourning and Violence*. London: Verso.

Clifford, James. 2009. 'Introduction: Partial Truths'. In *Writing Culture: The Poetics and Politics of Ethnography*, edited by James Clifford and George Marcus, 1–26. Berkeley: University of California Press.

Crehan, Kate. 2018. 'Antonio Gramsci: Towards an ethnographic Marxism'. *Anuac* 7 (2): 133–150.

Dauphinée, Elizabeth. 2010. 'The ethics of autoethnography'. *Review of International Studies* 36 (3): 799–818.

Jackson, Patrick. 2010. *The Conduct of Inquiry in International Relations: Philosophy of Science and Its Implications for the Study of World Politics*. London: Routledge.

Kanafani, Samar, and Zina Sawaf. 2017. 'Being, doing and knowing in the field: reflections on ethnographic practice in the Arab region'. *Contemporary Levant* 2 (1): 3–11.

Latour, Bruno. 2004. 'Why Has Critique Run out of Steam? From Matters of Fact to Matters of Concern'. *Critical Inquiry* 30 (2): 225–48.

Lorey, Isabell. 2015. *State of Insecurity: Government of the Precarious*. London and New York: Verso.

Nassif, Helena. 2017. 'On Punishability. Researching in Egypt after Regeni'. *Mada Masr*, February 3, 2017. https://madamasr.com/en/2017/02/03/opinion/u/on-punishability-researching-in-egypt-after-regeni/.

Marchart, Oliver. 2007. *Post-foundational Political Thought. Political Difference in Nancy, Lefort, Badiou and Laclau*. Edinburgh: Edinburgh University Press.

Palazzi, Franco, and Michela Pusterla. 2018. 'Remembering against the tide: Giulio Regeni and the transnational horizons of memory'. *Open Democracy*, January 25, 2018. https://www.opendemocracy.net/en/north-africa-west-asia/giulio-regeni-murder-transnational-memory-egypt-italy/.

Schatz, Edward. 2009. *Political ethnography: What immersion contributes to the study of politics*. Chicago: Chicago University Press.

Sjoberg, Laura. 2018. 'Failure and critique in critical security studies'. *Security Dialogue* 50 (1): 77–94.

Tsing, Anna. 2005. *Friction: An ethnography of global connection*. Princeton: Princeton University Press.

Van Wingerden, Enrike. 2017. 'Toward an Affirmative Critique of Abstraction in International Relations Theory'. *E-International Relations*, December 12, 2017. https://www.e-ir.info/2017/12/12/toward-an-affirmative-critique-of-abstraction-in-international-relations-theory/.

Wahba, Dina. 2018. 'Diaspora stories: Crippling fear and dreams of a better home'. *Mada Masr*, August 16, 2018. https://madamasr.com/en/2018/08/16/opinion/u/diaspora-stories-crippling-fear-and-dreams-of-a-better-home/.

Webster, Steven. 1986. 'Realism and Reification in the Ethnographic Genre'. *Critique of Anthropology* 6 (1): 39–62.

5

The Limits of Control? Conducting Fieldwork at the United Nations

HOLGER NIEMANN

Introduction

When I went to New York for the first time to conduct doctoral fieldwork at the United Nations (UN), I was expecting the unexpected. Using an interpretive framework in my Ph.D. project, I was aware that the process of 'diving into the field' (Yanow 2007, 116) implies changes, unexpected developments and last minute modifications of schedules and plans. Furthermore, I understood that I would arrive with preconceptions and expectations about my subject of study. Agreeing that 'there is no such thing as first contact' (Neumann and Neumann 2018, 1), I was aware of the situated construction of knowledge. Having read about it and having planned with this in mind, I was actually looking forward to facing the unexpected. And yet, when such situations occurred, for example when much anticipated interviews were cancelled or certain diplomatic practices of non-elected Council members turned out to be more important than I expected, it caught me by surprise. It took me a while to realise that this feeling of confusion – as uncomfortable as it felt at that time – was an important part of becoming sensitised to the situatedness of my knowledge of the field.

Fieldwork helped me to better understand the UN as my subject of study, but my insights were informed by a particular perspective. In order to get access to the UN, I formally gained the status of a civil society representative. While I was at no time pretending to actually represent a civil society organisation (CSO), it defined formally how the UN would treat me and significantly affected my opportunities to get access, gain insights, and interact with the field. Irrespective of all my careful planning and my reliance on a reflexive and interpretive methodology, the scope of this positionality surprised me. Ultimately, I did not want to miss out on experiencing the situatedness of my fieldwork knowledge. As a first-time field researcher, however, it made me wonder first and foremost about the limits of control of my fieldwork.

Of plans and realities

My fieldwork was part of a doctoral research project on the Security Council's primary responsibility for the maintenance of international peace and security. I was particularly interested in the role of justification as a practice of normative ordering (Niemann 2019). Defined as 'situated judgements' (Boltanski and Thévenot 2000), justifications are based on local and momentary references to knowledge, normative standards, and value claims. As such practices are not directly accessible, I used an interpretive framework, which considered knowledge to be situated in the local context of a research process (Wagenaar 2011, 23; Yanow 2006, 13). Interpretivism argues that this situatedness stems from an inextricable connection between the worlds of the knower and the known (Jackson 2011, 37). As the production of knowledge results from the researcher's embeddedness in his or her subject of study, this knowledge is inevitably an intersubjective and context-dependent construction. An interpretive research framework, thus, allows but also requires an open and reflexive research process for capturing the dynamics of this contextual construction of knowledge (Flick 2009, 20; Fine and Shulman 2009, 178). Therefore, I considered unexpected results not as failures of my research design, but as the very process of generating contextualised knowledge about my field. I also relied on the circularity of an interpretive research framework by taking into account a constant back and forth between the various steps of the research process (e.g. research design, operationalisation, data collection) (Yanow 2007, 118; Bueger and Gadinger 2014, 80). I took into account the need to constantly adapt my research process to the situation in the field, to be open to disruptions and changes as well as unexpected and surprising results. At the same time, my fieldwork also had to meet relatively practical requirements. These included my supervisors' expectations that I would return with 'proper' results, the funding agency's formal requirements that my proposed project outcomes were met, and ultimately my personal motivation to succeed in this endeavour. These practicalities created expectations regarding the conduct and outcome of my fieldwork as well.

When I arrived in New York, things turned out differently from what I had planned. Prior knowledge plays an important role in interpretive research as a driving force of knowledge production (Schwartz-Shea and Yanow 2012, 25).

I was conscious of the fact that the unexpected situations I found myself in represented textbook examples of how entering the field questioned prior knowledge. This is precisely what should happen when entering the field, as it generates the very 'condition for surprise' (Wagenaar 2011, 243) necessary from an interpretive perspective. Nevertheless, the scope of my situatedness was bewildering. It also made me realise the challenge of turning words into deeds when using an interpretive methodology. All of a sudden, my situatedness became a matter of questioning the success of my fieldwork, instead of providing a safeguard for ensuring the standards of my interpretive methodology were met. It took a while before I realised that this confusion is not only an inevitable part of the fieldwork experience (Wagenaar 2011, 245), but is ultimately also a source of unexpected and important insights into the UN and its practices. Uncomfortable as this situation was in the first place, this positionality was precisely the source of knowledge I was looking for.

A particular perspective on the UN

Elite institutions are known for being difficult to access and pose a challenge for fieldwork (Kuus 2013). This holds true for the UN as well. Sometimes scholars get access by becoming a delegate of a permanent mission (Schia 2013; Barnett 1997) or working within the UN bureaucracy (Mülli 2018). In my case, access was granted through accreditation with a CSO. The UN has a reputation for being often a shuttered place for CSO representatives. It is an intergovernmental organisation designed to serve the purposes of member states. Unlike in fields such as human rights or development, the field of peace and security is especially difficult to access for CSOs as consultation mechanisms are highly informal, decision-making processes often take place behind closed doors and CSO representatives experience a lack of transparency and public information. Knowing about this ahead of the fieldwork, I was aware of the limits of my status.

Furthermore, I was usually very clear on my role as an academic conducting research when interacting with people in the field. I considered this primarily a matter of formal status and was not especially interested in acting as a CSO representative. However, I was surprised by how much I became subject to the peculiarities of my formal status. In the following, I describe three different situations from which I gained specific insights due to my particular status when being in the field. In all three instances, I expected my initial plans to be challenged by the everyday realities. However, ultimately, they became much more important as they contributed to a particular understanding of my subject of study. It was only when reflecting on these situations that I realised that it was precisely in these moments that situated knowledge about the UN was generated.

The limited openness of the UN

The UN is first and foremost an intergovernmental organisation. It is constituted through its member states, which are represented in New York by delegates of their permanent missions. The UN is also an international bureaucracy consisting of a large body of staff. Civil society is neither formally part of the UN, nor is the UN – especially in the field of peace and security – particular well known for its openness towards CSO representatives. Nevertheless, CSO representatives play an important, yet informal, role in politics at the UN. They may act as agenda-setters, provide expertise, or lobby for stakeholders and because of that can be considered to be part of what has been labelled a 'third UN' next to delegates from UN member states and the UN bureaucracy (Weiss et al. 2009).

CSO representatives are frequently seen at the UN headquarters, interacting with delegates from UN member states and staff from the UN bureaucracy in various ways. At the same time, they have a particular kind of access to the UN, and especially to the Security Council. Numerous everyday instances reminded me of the differences between these three UNs and their distinctive, yet overlapping life worlds. My status allowed me to move around freely in the UN headquarters, and to access numerous official meetings of the UN and its various sub-bodies. However, I was only allowed to attend public Security Council meetings. While these meetings are important as formal sites for voting on agenda items, the literature often emphasises that the informal negotiations ahead of these meetings and meetings closed to the public, are much more important for the actual decision-making processes. Due to my status, I was unable to attend such meetings. However, being able to observe public Security Council meetings actually helped me in better understanding the importance of these meetings as a site for public justifications.

While I realised how much the UN is indeed an intergovernmental organisation that gives CSO representatives a different role, these public Council meetings also demonstrated to me how important the public nature of the justificatory claims made in the Security Council actually is. As claims of legitimacy or values, justifications can unfold their purpose in such settings, because their publicity provides the audiences necessarily needed for seeking recognition of justificatory practices by others. Justification can become relevant in the UN Security Council, because it can be explicated in public settings.

Concrete and blurring boundaries of the UN

Another moment of learning about the situatedness of my knowledge was when I realised the simultaneity of concrete and blurring boundaries in the

UN. At first sight, there is no doubt about the concrete nature of the UN's boundaries. The complex of the UN headquarters is clearly identifiable, not only due to the prominence of its buildings, but also because of the massive security architecture of fences, walls, and barriers surrounding it. Furthermore, entering the UN complex marks a clear line between inside and outside, with security checks and – depending on status – long queues to enter the site. I often physically experienced the fact that particular routes were closed to me due to my CSO affiliation. This held true, for example, for pathways through UN staff offices, elevators and areas with different designations. Often when I walked around the UN complex, I learned that my ID card, necessary to pass through electronic checkpoints and gates, prevented me without explanation from entering certain areas. Checkpoints determined my routes through the buildings, while security checks and the denial of access to particular parts of the UN headquarters became an important element of my daily transformation into a CSO representative in the eyes of the UN authorities when entering UN grounds.

Cleary identifiable boundaries also defined my position and my activities within the Security Council. The Security Council chamber has designated areas for the various audiences, not only defining different spaces, but also defining the competences of their respective occupants (Niemann 2019, 211). Delegates from UN member states, for example, are able to interact with Council members informally during meetings, while visitors are not permitted to move around the chamber, other than to leave via designated exits. Hence, interaction is prohibited and even access to information is limited. Papers and documents, which are sometimes circulated at short-notice in the chamber, are not available to the public in the visitors' gallery, with the implication that taking notes becomes crucial. I was literally trying to make sense of what was going on at the centre of the chamber by observing and taking notes.

While I could not directly speak with diplomats in these situations, I was able to observe how delegates interacted with each other and study the various practices diplomats performed in advance of Council meetings, to get an understanding of how social relations, the architecture of the Security Council chamber and diplomatic practices represented what has been called the 'dance' of multilateral diplomacy (Smith 2006). In correspondence with the architecture of the Council chamber, my formal status allowed me to look from a distance on this 'dance'. Unlike the diplomats actually performing the 'dance' at the centre of the Council chamber, my bird's-eye view allowed me to get a fuller perspective of their movements and how they interacted not only with each other, but also with the materialities of the Council chamber. Hence, my formal status defined not only where I was physically permitted to sit, talk, and walk, but also how I was able to conduct research. Coping with the material manifestations of the UN headquarters, but also how diplomats interacted with its symbols and objects, sensitised me for the manifestation of particular power relations and generated a certain type of knowledge.

At the same time, I learned the extent to which the UN is also a multi-sited field with blurring boundaries. Besides the permanent missions of UN members, an array of liaison offices from other international organisations and CSOs, think tanks and academic institutions that also work on and with the UN exist in close proximity to the UN headquarters. While wandering around, it struck me how much the entire neighbourhood at times actually constituted 'the field'. Unlike the UN complex with its strict protocol, these sites came with more unclear competences and mobilities. Security provisions in permanent missions, for example, varied considerably. Sometimes security clearances were difficult to obtain, with any electronic devices prohibited and checkpoints to be passed before I was taken to an anonymous meeting room. In other instances, I sat directly at the desk of my interviewee, often after walking through the offices of the permanent missions and getting an impression of the diverse ways they are organised. In addition, there were also a number of interspaces on the UN complex, such as floors, cafeterias, or the reading room of the Dag Hammarskjöld library, that were distinctive informal meeting zones. Here again the boundaries between inside and outside get distorted, as they provide spaces for encountering a diverse set of actors. These interspaces differed significantly in form and function, demonstrating the simultaneity of the concrete and blurring boundaries of my field.

Immersing myself in the rhythm of the UN

During my fieldwork, I was aware that 'being in the field' is a dynamic and often contingent activity. Timing is often more accidental than a matter of careful planning, with plans changing at short notice and the results of fieldwork subject to unpredictable developments. Therefore, I was expecting my schedules to be provisional. What surprised me, though, was the scope of becoming immersed in the particular rhythm of the UN. The UN calendar follows a specific course, which starts in September with the opening of the General Assembly. Due to funding formalities, my fieldwork partly fell in that period, which is usually not the best time for conducting fieldwork at the UN: Permanent missions are very busy preparing their governments' visits, the UN bureaucracy is completely focused on conducting the General Assembly, and the entire city is on alert due to the attendance of numerous heads of state. This affected, for example, the availability of interview partners, especially in smaller permanent missions, which were occupied with preparing for the General Assembly. At the same time, it helped me to better understand the annual life cycle of the UN.

Once I arrived at the UN, I became immersed in the daily schedule of the Security Council. The Council is by and large an ad-hoc body and much of its agenda is only planned provisionally. The monthly schedule, consisting of regular topics such as biannual discussions on topics such as the protection of civilians or mission mandate extension, is often complemented by meetings scheduled on short notice due to unforeseeable political circumstances. Thus, one of my everyday practices was to check the Journal of the United Nations, a daily publication that announces meetings and events at the UN. Nevertheless, my schedules were frequently subject to change, as high-profile diplomats in particular often had to cancel or relocate interviews at short-notice. Waiting, which often literally meant sitting in front of an office, became a frequent practice. At the same time, this waiting provided the occasion for reflection and writing-up memos or notes, and spending time in some of the interspaces, such as the UN cafeteria or semi-open floors, where I could attune myself to the rhythms of the UN at various times of day. Despite the frustration when interviews were rescheduled, the feeling of being driven by the heartbeat of the UN was significant in better understanding the UN's practices.

Explicating the partiality of fieldwork

Supposed fieldwork failures can lead to 'productive irritations' (Kurowska 2019, 85). The previous section gave insights into unexpected situations I faced during fieldwork and how my status allowed me to gain situated knowledge from such 'productive irritations'. Without overemphasising the impact of my formal status, it was an important source for getting a particular view on the UN. If I had a different role, my understanding of what the UN is and does would certainly have been a different one. Despite my attempts to be best prepared by reading numerous manuals and books on how to conduct fieldwork, these 'productive irritations' were confusing at first. Ultimately, however, this partial view on the UN was intriguing. It took a while for me to come to this conclusion, but it was important for better understanding how situated knowledge is actually constructed. Therefore, greater reflection, both on the challenges such confusing situations pose, and their potential for gaining insights seems advisable to me.

Obtaining access to the field, being able to generate data, establish networks and become acquainted with the field was important to me. As probably for every field researcher, I felt that 'not knowing is hard to tolerate' (Kurowska 2019, 76). Fieldwork, therefore, came with a certain interest in making sure telling things would actually happen when entering the field.

Although I knew that the realities in the field would not always meet my expectations, the scope of my positionality still surprised me. Ultimately, however, such situations became an important source of the kind of situated knowledge about my subject of study – knowledge that I was actually looking for. If fieldwork is precisely about that, then explicating our own situatedness and the partiality of the results it brings about is the first step in understanding supposed fieldwork failures as important sources of knowledge production. Since field researchers want to make the best of their work, they need to be opportunistic with those they talk to and what they observe (Fine and Shulman 2009, 179; Flick 2009, 226). This opportunism also generates the flexibility needed for adapting a research project to the complexities of the field. Being explicit about the opportunistic moves we made in the field, as much as about the limits in doing so, seems more honest with regard to the sources of our situated knowledge, and at the same time helps us understand not only the ways in which our field research occurs, but also how much our own situatedness is driven by questions of accessibility, power, and control. Explicating our own positionality points to the role of power relations in processes of knowledge construction (Haraway 1988), which is a longstanding issue in conceptualising fieldwork.

Fieldwork is a partial insight and it is necessary to reflect upon this partiality (Sjoberg 2019, 89). Therefore, we should strive for greater transparency in acknowledging the complexity of our fields, as well as the confusion our own situatedness can cause in the course of the research process. These are no trivialities, instead 'the limits of the art are part of the data' (Fine and Shulman 2009, 192). It remains to be discussed, though, how field researchers can actually achieve this. Some have argued that they should turn to autoethnography (Brigg and Bleiker 2010), others have pointed to the need for becoming aware of the underlying hegemonic discourses affecting the production of knowledge (Kurowska 2019). This particular volume talks about 'failures' and their importance for successful knowledge production. Irrespective of the avenue chosen, engaging in scholarly dialogue on the impact of situatedness, the limits and possibilities of particular roles in the field, as well as the confusion created by actually facing such situations seems vital for better understanding how situated knowledge is generated. It would allow us to develop different understandings of 'fieldwork failures' and reflect on how these challenges, as frustrating they may be at times, ultimately lead to precisely the surprising insights that motivate us to 'dive into the field'. There are no linear ways to cope with the challenges of fieldwork (Wagenaar 2011, 246). Talking and writing about how our research benefits from them – or not – seems an important, if not necessary, part of conducting fieldwork.

* *The author would like to thank Ann-Kathrin Benner, Caroline Kärger, Xymena Kurowska, the participants of the workshop 'On the Importance of*

Failure of Fieldwork: Living and Knowing in the Field' as well as the editors of this volume for their extremely helpful comments on previous versions of the manuscript.

References

Barnett, Michael. 1997. 'The UN Security Council, Indifference, and Genocide in Rwanda'. *Cultural Anthropology* 12 (4): 551–578.

Boltanski, Luc, and Laurent Thévenot. 2000. 'The Reality of Moral Expectations: A Sociology of Situated Judgement'. *Philosophical Explorations* 3 (3): 208–231.

Brigg, Morgan, and Roland Bleiker. 2010. 'Autoethnographic International Relations: Exploring the self as a source of knowledge'. *Review of International Studies* 36 (3): 779–798.

Bueger, Christian and Frank Gadinger. 2014. *International Practice Theory: New Perspectives*. Basingstoke: Palgrave Macmillan.

Fine, Gary A. and David Shulman. 2009. 'Lies From the Field: Ethical Issues in Organizational Ethnography'. In *Organizational Ethnography: Studying the Complexities of Everyday Life*, edited by Sierk Ybema, Dvora Yanow, Harry Wels and Frans Kamsteeg, 177–195. London: Sage.

Flick, Uwe. 2009. *An Introduction to Qualitative Research*. Los Angeles: Sage.

Haraway, Donna. 1988. 'Situated Knowledges: The Science Question in Feminism and the Privilege of Partial Perspective'. *Feminist Studies* 14 (3): 575–599.

Jackson, Patrick T. 2011. *The Conduct of Inquiry in International Relations: Philosophy of Science and Its Implications for the Study of World Politics*. London and New York: Routledge.

Kurowska, Xymena. 2019. 'When one door closes, another one opens?: The ways and byways of denied access, or a Central European liberal in fieldwork failure'. *Journal of Narrative Politics* 5 (2): 71–85.

Kuus, Merje. 2013. 'Foreign Policy and Ethnography: A Sceptical Intervention'. *Geopolitics* 18 (1): 115–131.

Mülli, Linda. 2018. 'Die Rituale der UNO? Wie habitualisierte Praktiken soziale Ordnungen und Hierarchien herstellen'. In *Ordnungen in Alltag und Gesellschaft: Empirisch kulturwissenschaftliche Perspektiven*, edited by Stefan Groth and Linda Mülli, 37–58. Würzburg: Königshausen & Neumann.

Neumann, Cecilie B., and Iver B. Neumann. 2018. *Power, Culture and Situated Research Methodology: Autobiography, Field, Text*. London: Palgrave Macmillan.

Niemann, Holger. 2019. *The Justification of Responsibility in the UN Security Council: Practices of Normative Ordering in International Relations*. London: Routledge.

Schia, Niels. 2013. 'Being Part of the Parade: "Going Native" in the UN Security Council'. *PoLAR: Political and Legal Anthropology Review* 36 (1): 138–156.

Schwartz-Shea, Peregrine, and Dvora Yanow. 2012. *Interpretive Research Design: Concepts and Processes*. New York: Routledge.

Sjoberg, Laura. 2019. 'Failure and critique in critical security studies'. *Security Dialogue* 50 (1): 77–94.

Smith, Courtney B. 2006. *Politics and Process at the United Nations: The Global Dance*. Boulder: Lynne Rienner.

Wagenaar, Hendrik. 2011. *Meaning in Action: Interpretation and Dialogue in Policy Analysis*. Armonk: M.E. Sharpe.

Weiss, Thomas G., Tatiana Carayannis, and Richard Jolly. 2009. 'The "Third" United Nations'. *Global Governance: A Review of Multilateralism and International Organizations* 15 (1): 123–142.

Yanow, Dvora. 2006. 'Thinking Interpretively: Philosophical Presuppositions and the Human Sciences'. In *Interpretation and Method: Empirical Research Methods and the Interpretive Turn*, edited by Dvora Yanow and Peregrine Schwartz-Shea, 5–26. Armonk: M.E. Sharpe.

Yanow, Dvora. 2007. 'Interpretation in Policy Analysis: On Methods and Practice'. Critical Policy Studies 1 (1): 110–122.

6

Tears and Laughter: Affective Failure and Mis/recognition in Feminist IR Research

LYDIA C. COLE

This discussion of fieldwork failure draws on my experience conducting research in Bosnia and Herzegovina (BiH) in 2015 as part of my doctoral studies which focused on the legacies of wartime sexual violence. Seeking to understand how complex subjects are produced by and produce themselves against post-conflict justice processes, this research located a series of intersecting frames of recognition. Inspired by Judith Butler (2006; 2009), these frames were visibility, legal-bureaucratic recognition, psychological recognition and witnessing. In this research, emotion and affect were central. Initially recorded in the margins of my fieldwork diary and spilling into conversations with friends and family over Skype, the practice of writing with and through emotion became both an ethical imperative and a way to more fully attend to the multiple inflections of recognition within post-conflict justice. In writing this chapter, I have been prompted to reflect on this process with regard to its successes and failures. Putting instances of tears and laughter into focus, I suggest that emotions, affect, and their failures structure the research encounter in significant, and often, productive ways. Tears and laughter are examined as *productive affective failures* that, on further reflection, enabled both renewed insight and an embodied knowledge of the research context.

The chapter re-examines two interviews which draw forth these aspects of emotion and affect in the research encounter. Both interviews take place in the broader context of the preparation of witnesses for war crimes trials, with each organisation occupying a different position in the post-conflict justice milieu. The first failure takes place during the second of two interviews with a psychotherapist based at Vive žene (Vive Women) in Tuzla. Here, I reflect on tears as a complex, affective response that enables further understanding of the complexities within the psychological governance framework. The second focuses on an interview with Žene žrtve rata (Women Victims of War), a prominent survivor association in Sarajevo. Here, I reflect on the perception of laughter in the research encounter, focusing on how this provoked a moment of disconnect and reoriented the interview toward renewed understanding. In each case, I reflect on my interview transcripts, the writing of these encounters in my fieldwork diaries, and my process of writing emotions in, and out, of the research encounter and in my doctoral thesis.

Emotion, affect, method

Before I began my research, in 2011, a forum featuring several prominent feminist IR scholars interrogated 'the question of *whether* and *how* emotions should enter our scholarship' (Sylvester et al. 2011, 687). While feminist scholars have long advocated for emotions to be taken seriously as constitutive of the political, social, and cultural world, the forum specifically sought to address the ways that emotion was still being written out of the research encounter. Sandra Marshall, a contributor to the forum, reflected on her feelings of 'awe' listening to these 'super-human' feminist researchers who were able to speak of their experiences with 'unwavering composure' (Ibid., 688–9). Later, on reflection, 'feminist alarm-bells . . . started ringing'. Exploring a 'culture of silence' surrounding researchers' emotions within the discipline (Ibid., 689), Marshall sought to 'uncover some of the untold stories' about emotion (here, specifically trauma) in feminist international relations research (Ibid., 690). This forum, and feminist and critical methodological interventions, especially those that emphasised the personal, the emotional, and the affective (e.g. Daigle 2015; Dauphinée 2007), as well as the inherent relationality of the research encounter (Stern 2005), were instrumental as I prepared for my fieldwork. These texts provided key insights into the way that emotion and affect, both my own and those of my participants, would structure my research and its frames of recognition. Coming to fieldwork prepared for emotions to enter my research, I nevertheless found myself unprepared for precisely how and to what extent.

The term productive affective failure is inspired by the broader literature on affect and failure in feminist international relations research. It draws specifically on the concept of 'affective dissonance', applying this to specific moments of fieldwork failure. Linda Åhäll and Clare Hemmings both frame affective dissonance as a starting point for feminist inquiry. For Hemmings (2012, 154), this concept of affective dissonance is proposed as a 'critique of empathy as the basis of an affective feminist solidarity'. Grounding through a narration of her process of becoming a feminist, affective dissonance is a process of identity formation which arises from a dissonance between a 'sense of self and the possibilities for its expression and validation'. This is a basis for – though does not necessarily always lead to – 'a connection to others', a 'desire for transformation', and forms of solidarity which are nevertheless 'thoroughly cognisant of power and privilege' (Ibid.). My reading is closer to Åhäll's (2018, 44) reformulation which pinpoints the concept as a 'methodological tool for analyses of the politics of emotion more broadly'. Putting affective dissonance to work in research, and in fieldwork specifically, entails an openness to being transformed in the research encounter in ways that generate new insights and embodied knowledge that would not have otherwise been possible.

I focus here on instances of affective dissonance that nevertheless, at the time, felt like failure. Laura Sjoberg's reflections on failure and critique in critical security studies are instructive. Contending that 'failure should be recognised and embraced rather than ignored, covered up, or compensated for' (2019, 77), Sjoberg situates failure as a 'crucial part of the practice of critique rather than a shameful secret and an embarrassing shortcoming' (Ibid., 89). The term failure is then used deliberately. The affective responses discussed – tears and laughter – can be understood as failure to the extent that they run contra to myths of an unencumbered, unemotional researcher. Further, in context, they were affective responses that were unexpected, even inappropriate. However, and importantly, even as they structured what I knew and wrote about the frames of post-conflict justice in BiH, in the process of writing, I often obscured these affective responses from the research encounter. Productive affective failure can be understood as a subset of affective dissonance. The term helps me to examine the paradoxical nature of research encounters that seem like failure, while nevertheless becoming central to a renewed knowledge of oneself in the research context. During my research, there were likely other instances of affective failure that I have forgotten or perhaps did not even register at the time. Here, I choose to concentrate on affective failures – tears and laughter – that produced insights for my research and instances where this failure seemed to facilitate further understanding, if not connection to those with whom I spoke.

Before turning to tears and laughter, it is worth briefly outlining some practicalities of the research approach. The oft-cited, and sometimes palpable 'research fatigue' in BiH,[1] impacted the way that I engaged with research participants (Clark 2008). Those with whom I spoke had varying expectations of interviews, yet, almost all were used to speaking with researchers and journalists. Though coming to interviews with a commitment to narrative

1 In my wider research, this was an important factor that was brought to bear on the analysis of the frames of post-conflict recognition.

research, it was often difficult to discuss issues which deviated from publicly available materials. Given this, I adopted an approach to interviews that I describe as (semi)unstructured. Coming to interviews prepared with topics and questions and sending these to participants where they were requested, I nevertheless emphasised my interest in them as people – attending to the specificities of their role and their thoughts about the topics of discussion. I tried to curate a sense of familiarity by sharing my own thoughts and experiences. For example, I often drew comparisons with the UK context, highlighting key issues and limitations in policy and practice with regard to relevant aspects of gender and welfare. Both interviews discussed took this tack, with varying degrees of success and, indeed, failure. All participants were given the option of conducting interviews in English or Bosnian. For interviews conducted in Bosnian, I worked with a trusted translator who had experience with the questions that informed the research. Interviews discussed in the section on tears were conducted in English since the psychotherapist had a good working knowledge of English, while the interview discussed in the second on laughter with the representatives of Žene žrtve rata was conducted in Bosnian. Working with a translator enabled insight into the nuances of language and expression that would have otherwise been missed, yet, it also created a sense of distance between myself and those of whom I was asking questions. Indeed, it was a contributing factor to the intersections of failure and (dis)connection that are discussed in this chapter.

Tears: On the couch at Vive žene

Vive žene was founded in 1994 with the support of a women's group in Dortmund, Germany in response to the violence that accompanied the war. The organisation aimed to provide ongoing support and psychosocial care to women and children who had experienced a range of war-related violence, including torture, displacement, and sexual violence. Over the course of 25 years the organisation has built on this work, while adapting to a changing post-war context. Continuing to work with victims of war-related violence and displaced communities, they have expanded their remit to include other forms of violence. Drawing on this experience, the organisation actively works within the post-conflict and transitional justice context, providing training to legal professionals and other non-governmental organisations and conducting advocacy work related to the individuals and communities that they work with.

Over the course of my fieldwork, I met with a psychotherapist working for Vive žene on two occasions – in April and October 2015. On both occasions, I travelled to meet her at their office in Tuzla. Both conversations took place in the bright and airy therapeutic rooms in the building, putting me at ease almost instantly. Our first conversation was informative, centring on the organisation's work with victims of war-related violence and its psychosocial approach. Our second conversation was more focused. In the preceding months, I had come to reflect on the intersecting layers of post-conflict justice. I noted the way that the therapeutic relationship and the language of psychological recognition were deployed within legal-bureaucratic frames of justice. Resonating with conceptual discussions on the (dis)connections between trauma, truth-telling, catharsis, and the law (e.g. Minow 1998; Moon 2009), I began to map this intersection empirically. I became interested in understanding how psychosocial organisations negotiated these in their work, asking questions about the psychological preparation of witnesses for war crimes trials.

During the second interview, we came to this topic through a broader discussion about the role of the therapeutic encounter in facilitating clients to speak about traumatic experiences. Talking, remembering, naming and listening was 'part of the process [... of] "healing for trauma", but the psychotherapist emphasised that this might also 'be preparation for witnessing'. A much,

> [S]maller number [...] of our clients, they decide after psychotherapy to become a witness [...]. And for witnesses especially, it is very important to go through psychotherapy. .To become more stable, to really have the feeling that you are in control. That you know everything that has happened.

(Anonymised, 2015a)

Continuing, the psychotherapist explained that it was their role 'to go with the clients to be there, to prepare them, but not then go in the court and be a witness'. Briefly interjecting, I asked whether it was common for those working in the organisation to be called as expert witnesses during war crimes trials.

> **Psychotherapist**: No, it was sometimes. It was when they were judging persons like Biljana Plavšić[2], when there was some person, and they needed some proof that really things have happened. And so, they were gathering all sorts of

---

2 Plavšić served as a member of the acting Presidency of the so-called "Serbian Republic of Bosnia and Herzegovina" (later Republika Srpska). In 2001, she was indicted by the International Criminal Tribunal for the former Yugoslavia (ICTY) for genocide, war crimes and crimes against humanity. Plavšić later pleaded guilty to persecutions, a crime against humanity, and was sentenced to 11 years' imprisonment (see Plavšić (IT-00-39 and 40/1)).

experts [...] What was happening and what were the consequences for the clients?

Author: But that only happens in specific cases when they're trying to establish a wider picture – so perhaps expert witnesses would be required for someone who was higher up the chain of command […]. But in a more localised sense, it would just be enough for witnesses to be called for people who experienced violence.

Psychotherapist: Yes, yes.

Author: Sorry, this is always such a heavy topic to discuss.

Psychotherapist: Yes.

[Silence]

Author: Yes, sorry.

Psychotherapist: You are crying, why? How does that affect you? What's happening?

Author: It's okay, I'm okay.

Psychotherapist: But why are you crying?

Author: I think sometimes it just quite overwhelming to speak about. I'll be okay.

Psychotherapist: Okay.

Author: Maybe we can try a different question.

(Ibid.)

The recording of the interview continues for around ten minutes, during which the psychotherapist and I discussed several other topics including her opinions of a prominent feminist truth-telling process coordinated by Žene u crnom (Women in Black). With the interview coming to a natural pause, I turned off the Dictaphone. At this point, the psychotherapist again expressed her concern. Offering to make coffee, we continued to talk.

In my fieldwork notes after the interview, I wrote at length in an attempt to process what had happened.

13 October 2015

> I just had a meeting [with the psychotherapist] which was quite uncomfortable. At some point in the interview I felt the need to cry. I think [the psychotherapist] located this in some trauma I have. I'm not sure if I have a trauma [...] our conversation today seemed to transcend the researcher-researched boundary. When I was upset, she asked me what was wrong; did I have some connection to Bosnia? Was it something about the Court process? I replied that I was tired [...] a lot of [this research is ...] emotionally hard work. I forgot what we talked about when I turned off the recorder.

Further, and actively reflecting on how and whether I would write this into the research encounter, I wrote:

> [W]ill I write about this? Should I write about how [it] interrupted for a moment the way the interview went. She became and was reinforced as therapist. She asked me questions and made me coffee. How did it feel to be her patient?

The interpretation of the interview in the final version of the thesis was more muted. Though the interview is subject to discussion, my reflections on the tears were confined to the margins. The interview is mentioned in the introduction within a discussion of methodology and co-production of narrative during interviews. Particularly, I discuss my approach to follow-up interviews where I made space for 'participants to respond to, push back against, and develop' my interpretation of our previous discussions (Cole 2018, 25). I re-approached these conversations with the psychotherapist more directly in the conclusion, reflecting that I had come 'to empathise strongly with the complex negotiations that psychological professionals made regarding the post-conflict justice context'. Describing the intersubjectivity of the encounter, and placing this in a broader context of post-conflict psychological recognition I wrote that,

> [T]his empathic relation was not unidirectional. In this context, my questions regarding the subject of wartime sexual violence were often turned back toward me. Throughout the interview, and beyond the context of my research statement, the psychotherapist wanted to know what my interest in the topic was, and in what ways did it affect me. During this interview, as we continued to talk about the work of the organisation, the power relations in the conversation were ever-shifting. As I came to recognise the negotiations that the psychotherapist made through her work with clients, I was placed 'on the couch', layers of psychic and social recognition moving between us in the encounter.
>
> (Ibid., 277)

The interview was key to my reading of psychosocial organisations within the contemporary post-conflict justice context, with several identifiable and structuring effects on my research. First, it prompted an examination of trauma which moved beyond Foucauldian-inspired IR literature that explores how it is mobilised as a tool of governance (Howell 2011; Pupavac 2001). Unsettled by the feeling of connection to the psychotherapist, I was compelled to write in a way that conveyed the affective complexity within the encounter and in a manner that better reflected the way that organisations like Vive žene mediate the relationship between the psychological and the legal for their clients. Second, and perhaps paradoxically, my reaction of tears within the interview inaugurated a re-examination of the difficulties inherent in speaking about trauma. In my (very limited) experience of being placed on the couch, I had resisted the attempt to pinpoint my tears in a categorical or defined manner. This experience – though clearly irreducible to those who have experienced war-related harms – nevertheless enabled an embodied insight into the possibilities, limitations, and (potential) violences of narrating trauma through the various structuring frames of post-conflict recognition. These insights were invaluable as I examined the role of testimony and witnessing in post-conflict recognition. The next section reflects further on witnessing and mis/recognition, discussing laughter during an interview with two representatives of a Sarajevo-based survivor association.

Laughter and mis/recognition

In November 2015, I conducted an interview with two representatives from Žene žrtve rata. Founded in 2003, two years after the first successful prosecutions of rape as a crime against humanity in the International Criminal Tribunal for former Yugoslavia in the Foća trial (Helms 2013, 197), the organisation aimed to offer support and aid to survivors. Since then, the organisation's director Bakira Hasečić had become a highly visible and vocal figure, both within BiH and internationally. Positioned as a public advocate for survivors of wartime sexual violence, Bakira Hasečić and the organisation more broadly have become known for their work facilitating the prosecution of war crimes (Ibid., 213). When I met with the representatives, I hoped to gain a better understanding of their role with regard to the various legal and bureaucratic institutions which pursue war crimes prosecutions. In my preparatory notes, I wrote that the 'association gathers evidence and information about war crimes [...] with a view to prosecution' and that the organisation had previously 'provided key testimony in rape and sexual abuse trials' and had 'helped obtain justice – financial and psychological for many of its [...] members'. Reframing these notes for the interview, I noted the following guiding questions:

- In what ways are you able to help members of the organisation obtain justice?
- Have you been satisfied with court processes to date?
- What more could be done for women victims of war?

When I arrived at the office – located at the bottom of a residential building in Sarajevo – it became clear that I had come at a difficult time. Joined a few moments later by my translator, we were asked to take a seat and wait. Soon after, we were called through to one of the rooms at the back of the office. As we sat down, one of the representatives intimated that the identity of a protected witness had been revealed.

Despite this, the interview began quickly. Crowded around a small table in the office, the representatives seemed keen to tell me about the current work of the organisation, including the publication of a monograph which detailed the extent of wartime sexual violence. The long informational statements given by the representatives took some time to translate. During this process, both representatives left to take a phone call. After a brief intermission, one of the representatives returned and recommenced the interview. Despite the confusion that seemed to dwell in the gaps in conversation, the first part of the interview covered much ground in terms of the organisation's commitment to 'break the silence' surrounding wartime sexual violence in BiH. As the interview progressed, I attempted to direct conversation toward other aspects of post-conflict support and recognition. In doing so, I hoped to prompt further reflection on their positioning with regard to the frames of post-conflict recognition that I was beginning to identify.

At this point, both representatives had returned to the room. While the first representative remained seated around the table with the translator and I, the second representative had positioned herself outside of this circle behind a desk near the office window. Adjusting to the new dynamics of the room, I asked:

> **Author**: So [...] we've spoken quite a lot about the ways in which you help women achieve justice and some of the psychological support. But it also mentioned on your website that you also help with economic issues of women. And I was wondering whether this was to do with the civilian status of war category? [...] Or whether it extended beyond that?
>
> **Representative 1**: When it comes to the economic support of women, it is also conducted through different projects. So, it all depends on the availability of projects and different propositions where we can supply with the projects. And when it comes to the status of civilian war victims, it has been introduced to the law, and I think at this point they receive 586 marks per month. Is this it?
>
> **Author**: So, maybe we can return a little bit to the book. Is that okay?
>
> **Representative 1**: I don't know what, [Anonymised] has suggested she has been working on, and so...
>
> **Author**: I just...
>
> **Representative 2**: We have given you the brochure and I think it is enough for your project. You have everything in that – all the information.
>
> **Translator**: So, I think we're done.
>
> **Author**: Okay, I think that's it. Okay. Thank you.
>
> **Representative 2**: Were you laughing?
>
> **Translator**: They think we were smiling. But I didn't...
>
> **Representative 2**: We have given you enough information.

> And you have all the information in the brochure. ... Because this is not the first time that we helped the students who worked on their thesis.
>
> **Author**: Okay.
>
> **Representative 2**: And we're always available for it.
>
> **Author**: Great, thank you very much.
>
> (Anonymised 2015b)

Reconstructing from fieldwork diary notes and my recollections of the incident, the recording of the interview ends abruptly, and the translator and I are shuffled out of the office. Yet, the conversation did not end there. With the second representative acknowledging a shift in tone, the tension in the room all but dissipated. As we clutched jackets and bags, the second representative addressed me again. Expressing regret for the way the conversation ended, she intimated a lack of trust in researchers: though many came to do interviews with the organisation, this rarely translated into outcomes. It neither changed the position of survivors in BiH, nor gave the organisation anything tangible to build into their advocacy work. After the interview, the translator and I wandered slowly back to the city to process the encounter.

This interview underlines a more fundamental disconnect between the representatives and me, in my role as researcher. The representative's accusation of misrecognition – through a smile or laughter – reveals important complexities in terms of witnessing in the aftermath of harm. Though from the outset, the representatives were clear about their desire to 'break the silence' around wartime sexual violence, this was not all that I was called to account for in our further conversation. Disconnection opened space for the representatives to return my questioning. Elaborating on past experiences speaking with students, researchers, and journalists, she intimated that they were tired of answering the same questions. In doing so, she called me to account for a longer trajectory of extractive knowledge production. Interrogating me in my position as researcher, the representatives wanted to know how *this* interview and *this* research project could be any different. In the context of my thinking on testimony and witness, this encounter prompted reflection on the multiple and intersecting forms of harm that those affected by war-related violence might call researchers to account for. Specifically, these harms were not temporally defined by war or conflict; rather, they evolved and took on new meanings over time and through encounter.

Laughter, as well as a process of dwelling on the affective elements within the disconnect, opened a means through which to explore the ambiguities of recognition for the subject of wartime sexual violence; and prompted further examination of my complicity in the reproduction of post-conflict harms.

Conclusion: Taking note of affective failure

This chapter put into focus two instances of fieldwork failure, engendering discussion of its productive elements. While instances of tears and laughter provoked differing responses and feelings of dis/connection, they are drawn together to demonstrate how affective failure, and its subsequent reflection, can lead to a transformed, embodied knowledge of the research context; in this case post-conflict recognition in BiH. Interviews with the psychotherapist at Vive žene revealed a complex negotiation of psychosocial recognition across therapeutic and legal contexts. Tears disrupted power relations within the encounter, with the psychotherapist placing me 'on the couch'. While bringing to the fore an embodied (if still limited) understanding of the limitations and potential violences of the expectations and frames surrounding the narration of trauma, my feelings of dis/connection also prompted a more sympathetic reading of the psychotherapist's position. The encounter at Žene žrtve rata similarly revealed a complex interplay between forms of recognition. Laughter was disruptive to the extent that it enabled the representatives to ask questions of me, holding me to account in my position as researcher. The accusation of laughter, along with subsequent discussions, enabled an embodied reflection on both my preconceptions of what it meant to bear witness in the aftermath of war-related harm and my own potential for complicity in the reproduction of this harm. Tears and laughter can be understood as prompts toward introspection on the role of power, positionality, and hierarchy within fieldwork; specifically, in the context of post-conflict recognition. However, to be a productive force in the context of research, this process of introspection must be learned from and put to use. In the context of my research, affective failure enabled a more reflexive approach to unfolding the multiple inflections of recognition in the context of post-conflict justice processes in BiH, and one which was more attuned to the complexities of power and vulnerability within encounters.

Affective failure is bound to happen during research. Contra to the myth of the encumbered, non-emotional researcher, a plethora of affective responses – both conscious and unconscious – enter our research frames. Feminist, critical, and ethnographic-style research tends to open up these questions of emotion, affect, and response. Taking seriously aspects of body language, relationships with research participants and translators, feelings of dis/connection, and reflections on positionality and hierarchy, these concerns are placed at the centre of the research. This chapter cannot tell you precisely how or to what extent affect and failure will enter your research. However, it might reassure you to know that affective failure is not a failure of research practice. Rather, it is another potential site for learning and unlearning our preconceptions, experiences, and training. And an important part of the process of continued reflection on our topics of research and of ourselves.

References

Åhäll, Linda. 2018. 'Affect as Methodology: Feminism and the Politics of Emotion'. *International Political Sociology* 12 (1): 36–52.

Anonymised. 2015a. 'Interview with Psychotherapist at Vive žene, Tuzla.' Interviewed by Author. October 12, 2015. Audio, 1:00:10.

Anonymised. 2015b. 'Interview with Senior Representative and Project Assistant at Žene žrtve rata, Sarajevo'. Interviewed by Author. November 16, 2015. Audio, 48:43.

Cole, Lydia. 2018. *The Subject of Wartime Sexual Violence: Post-Conflict Recognition in Bosnia and Herzegovina*. Ph.D. diss., Aberystwyth University.

Butler, Judith. 2006. *Precarious Life: The Powers of Mourning and Violence*. London: Verso.

Butler, Judith. 2009. *Frames of War: When is Life Grievable?* London: Verso.

Clark, Tom. 2008. '"We're Over-Researched Here!" Exploring Accounts of Research Fatigue within Qualitative Research Engagements.' *Sociology* 42 (5): 953–970.

Daigle, Megan. 2015. From Cuba with Love: Sex and Money in the Twenty-First Century. Oakland: University of California Press.

Dauphinée, Elizabeth. 2007. *The Ethics of Researching War: Looking for Bosnia*. Manchester: Manchester University Press.

Helms, Elissa. 2013. *Innocence and Victimhood: Gender, Nation, and Women's Activism in Postwar Bosnia-Herzegovina*. Madison: University of Wisconsin Press.

Hemmings, Clare. 2012. 'Affective Solidarity: Feminist Reflexivity and Political Transformation'. *Feminist Theory* 13 (2): 147–161.

Howell, Alison. 2011. *Madness in International Relations: Psychology, Security, and the Global Governance of Mental Health*. Oxon: Routledge.

Minow, Martha. 1998. *Between Vengeance and Forgiveness: Facing History After Genocide and Mass Violence*. Boston: Beacon Press.

Moon, Claire. 2009. 'Healing Past Violence: Traumatic Assumptions and Therapeutic Interventions in War and Reconciliation'. *Journal of Human Rights* 8(1): 71–91.

Pupavac, Vanessa. 2001. 'Therapeutic Governance: Psycho-social intervention and trauma risk management'. *Disasters*. 25 (4): 358–372.

Sjoberg, Laura. 2019. 'Failure and Critique in Critical Security Studies'. *Security Dialogue* 50 (1): 77–94.

Stern, Maria. 2005. *Naming Security – Constructing Identity: 'Mayan-women' in Guatemala on the Eve of Peace*. Manchester: Manchester University Press.

Sylvester, Christine et al. 2011. 'The Forum: Emotion and the Feminist IR Researcher'. *International Studies Review* 13 (4): 687–708.

Part III

Understanding and Connecting

7

The Valorisation of Intimacy: How to Make Sense of Disdain, Distance and 'data'

EMMA MC CLUSKEY

Scene one: Fez

When I arrived in Fez, my Moroccan contact and so-called 'gatekeeper' granting me access to the migrant camp I was due to visit, told me that there had just been a public flogging there. Sarah explained, very matter-of-factly, that one of the migrants living in the camp, a Nigerian man, had been found guilty in a makeshift 'court' of stealing, and was thus sentenced to 30 lashes.

'Marginalised communities living outside the law tend to always have their own forms of justice', Sarah explained to me whilst driving me back to my guesthouse, and told me a story about a particular group of Berbers who punish stealing of water with a requirement that that person then cooks dinner for the whole community. It wasn't exactly the same thing. 'There are public floggings all the time in that camp'. As the director of a migrant NGO in Morocco for almost 20 years, she seemed to be hardened to all this. I however felt immediately sick; horrified and appalled. I hadn't yet been to the camp or met any of the people living there; all from sub-Saharan African countries, mostly male. But I wondered what kind of people could inflict this kind of pain on somebody else in such a barbaric and calculated way. I was supposed to be broadly researching the effects of European bordering technologies on so-called third country nationals attempting to reach Europe. Like many others working within the field of security and migration, especially during the most recent refugee 'crisis', my main *problematique* was the complete de-humanisation of people on the move which has systematically taken place over the past 30 or so years; a misery which was very much a rule of the game rather than an exception in times of 'crisis' (Jeandesboz and Pallister-Wilkins 2016). Our project had investigated how various technologies had impacted the journeys of these 'TCNs' (Third Country Nationals), removing freedom of movement from the equation and reconceiving security as a 'balance' between coercion and surveillance. People crossing borders, with all their myriad of stories, families, careers and desires were homogenised and funnelled into flashing dots on a screen, FRONTEX statistics or a racialised horde blocked by police, barbed wire and dogs. An analysis informed by anthropology was seen as a way to re-humanise these travellers, disrupting statist or bureaucratic accounts of migration so prevalent in International Relations by privileging fragmentation and journeys, both temporal and spatial (see Basaran and Guild 2017; Bigo and Mc Cluskey 2017).

At first, it felt like a slightly futile endeavour, something we were forced to insert into a big European Commission-funded project to avoid a sterile or technocratic narrative of border technologies, nowadays often absurdly framed as 'humanitarian' (Gabrielsen Jumbert 2013; Pallister-Wilkins 2015). I was the ethnographer on the project, so it was me who could fly off and produce this research, with these refugees as my interlocutors (in reality, as with many of these types of projects, there was not really a great deal of time for in-depth participant observation or deep 'hanging out' [Madison 2005] so 'ethnography' became 'ethnographic interviews').

I knew it would be difficult; confusing emotions and feeling continuously destabilised are all part and parcel of fieldwork. It is always hard to speak about 'migrants' experiences' without replicating the usual discourses of vulnerability and suffering. This is especially the case when these individuals are forced to exist in an enclosed space, exposed to journalists and academics turning up to 'research' their daily lives (see Picozza 2018). But I had already formed a tentative opinion about the people in the camp and it left me conflicted: they were victims of these bordering regimes, certainly. But they also seemed frightening and violent. I'd never carried out fieldwork with people I had felt such antipathy towards before. Especially when my specific role was to 're-humanise' them in some way or another.

So, the following day, I made my way to the informal camp to introduce myself. As I knew beforehand, the camp was divided according to nationality, with different roles allocated to different national groups. Nobody was particularly welcoming, but then again, I wasn't particularly friendly either. I wasn't expecting to be greeted warmly or offered a drink. But the wariness of my first interviewee was surprising, nonetheless. His words, 'You are making a living from our suffering' haunted me for weeks. It was true. I was. Sub-Saharans in Morocco (for they were always presented as, and presented themselves as, a homogenous group) were over-researched already. It wasn't just 'an overcrowded field' (Andersson 2014), but profoundly sad and hopeless. I just didn't have it in me to try.

This was a failure from the outset. How could I even begin to attempt to do justice to the lived experiences of these people, trapped in Morocco, whose dreams and desires for Europe were funnelled and channelled into more militant claims for rights, counter-insurgency rhetoric and disdain towards me? Surely to talk about this, however sensitively, would feed into the securitarian and far-right narrative of civilisational clashes?

How could I talk about my own positioning, the discord I felt, the fear? Should I 'write this in?' How would this at all help in re-humanising these people who were already de-humanised in so many ways? The cognitive dissonance left me feeling rather depressed and I'm quite ashamed to say that I cut my losses and left after only a week.

Because of these conflicting feelings, I re-located my fieldwork from Fez to Rabat, where I also had an NGO contact. The capital city was also home to around 2000 Levantine Arabs on their way to Spain and further onwards into Europe. It would be much easier for me to build some kind of relations with people in Rabat as the situation for these Arab migrants, I had been told, was not so desperate or harrowing. Their suffering was less visceral. Stan Cohen (2001) has written in *States of Denial: Knowing About Atrocity and Suffering* about the under-theorised notion of compassion fatigue which can contribute to a denialist mindset. A psychological term which alludes to the sheer quantity and intensity of suffering leading to a type of numbing and de-sensitivity towards people enduring this suffering, 'compassion fatigue' is perhaps one explanation for my disengagement with the migrants in Fez. The strategies of avoidance Cohen speaks of are certainly recognisable in my own practices. However, this explanation lends a veneer of righteousness to what was a rather selfish action (indeed, the term is often deployed in relation to humanitarians who have spent too much time in 'the field'). I was no humanitarian, just someone on a 'jet-set' ethnography (Olwig 2007, 22) who would simply have rather hung out with people who made me feel less guilty and less uncomfortable from the outset.

A more academically acceptable reason for me shifting sites also presented itself, allowing me to leave aside these feelings of shame and disdain. There had not really been that much written about Arab refugees in Morocco as the phenomenon was rather new; a consequence of the 'way' to Europe being made much longer and pushed much further west as the result of various EU-Turkey, EU-Libya deals and increasing surveillance of the Eastern and Central Mediterranean routes (Frontex 2018; Heck and Hess 2017). I'd spent a bit of time in Syria before the war, worked with Syrian refugees during my Ph.D. and didn't even speak French, the language of many of the sub-Saharan people living in Fez.

I became rather sheepish when I was chastised for my decision to move to Rabat and interview Arabs by the chair of the NGO: 'For God's sake, Emma. Why are you interested in Syrians? They don't have a difficult time here at all compared to the Africans. They're the brother Arabs; they get given everything on a plate'. The deserving/non-deserving migrant dichotomy invoked by Sarah was a little surprising, but in some ways, she was right of course. But by that point I'd already made up my mind.

Professional ethics of empathy: Problems and pitfalls

Placing ontological primacy on lived experiences and all their complexities is central to any mode of anthropological inquiry. Our professional ethic demands us to be open and sympathetic to all our research subjects and the sets of the relations within which they are embedded. This is so well known that it's almost a doxa of the field. Ethnographers are supposed to be moral relativists. Not only does this prevent against 'conceptual enclosure' (Montesinos Coleman 2015), it can also shed light on the conditions of possibility for all the political 'bads' that have happened lately: Brexit, Trump and the like; in short, what Geertz (1984, 275) describes as 'looking into dragons, not domesticating or abominating them'. An openness and understanding of the experiences of everyone in the field, no matter how morally ambiguous this turns out to be – is supposed to be essential. But of course, this commitment to professional empathy is much easier to grasp in the abstract. When one is immersed in distressing or destabilising situations, you can never predict exactly with whom bonds will be formed. Professional ethics can only take you so far.

Relatively little has been written about the problems encountered in the field thanks to this professional commitment to empathy. In the field of IR, Julian Eckl (2008) was one of the first to question what 'responsible scholarship' looks like when researchers find themselves in ethically ambiguous situations – concerns which also relate to what and how researchers should convey their findings. From a gender perspective, 'awkward surplus' is the phrase used to talk about the stuff that gets written out of the final book or paper – the fear of what informants come to expect in return within the very mercenary relationship of the researcher and the researched, for example

(Hanson and Richards 2017). As women in the field, what amount of sexual harassment will we put up with to form a relationship with those we are researching (see also Desirée Poets's chapter in this volume)? Though I experienced no harassment at the hands of my interlocutors in Morocco, the 'awkward surplus' that manifested itself in my fieldwork also seems to be taboo. They disliked me; I kept them at arm's length. For my interlocutors to view me in any way other than as an annoyance would have seemed like a perverse improvement of sorts at the time.

These types of dilemmas and contradictions arise because of the value the discipline places on forming *intimacies* and intimate relations with research subjects (Appadurai 1997). As Hanson and Richards (2017, 596) have remarked 'No one gets excited about an ethnographer that has awkward or strange relationships with the communities they are trying to work in'. Certainly, the 'best' ethnographies, the ones which are most celebrated, cited and win all sorts of awards – are the ones in which close bonds have been formed.

There is no doubt that the level of discomfort I felt shaped my research in Morocco. But to dwell too much in this discomfort would have felt self-indulgent, possibly somewhat racist, and perhaps insignificant in the eyes of other, more experienced scholars who might be more hardened to this kind of violence and suffering. I didn't want to be seen as too soft to do this type of research; naïve and immature, clueless as to the horror of the suffering of refugees at Europe's borders, even though that was all probably true. This discomfort also didn't fit the line of the big research project. Instead of acknowledging this therefore, I just pretended it didn't happen. It was relegated to the stuff of half-drunken conversations with close colleagues at conferences.

Scene two: Rabat

The *Rue al-Arab al-Maghriby* was temporary home to many of the Syrians and Yemenis passing through Morocco on their way to Europe. I was introduced to Samar in a Syrian restaurant over a *narghile* and orange juice. A young, articulate and enthusiastic woman who was very easy to chat with, I had met her much more serious husband a few days earlier and was heartened by the high esteem in which he had obviously held his wife. Both of them were extremely positive about my research: 'It's great that you will write about us and bring attention to our stories. Syrian people here have had to travel through six, seven countries and families are having to storm the wall (in Ceuta), with their children, everything! We are a civilised people.'

Rabat was a much more pleasant fieldwork experience altogether. There were no 'camps' here. The travellers who were here were a bit more settled, living in cheap hotel rooms or cramped apartments. There were NGOs dealing with the large number of migrants who found themselves in the city, as well as a UNHCR presence. For many of the Levantine Arabs however, 'official' NGOs were avoided in favour of more informal Syrian diaspora organisations which had a long history in the city. Keeping under the radar was often a way of life for many of the Syrians and suspicion of charitable organisations, even those that could have offered them some accommodation or small payments, meant that they were much more difficult to locate. I was quite pleased with myself for being able to gain the trust of many of the Arab travellers.

Samar's story was very relatable – she was a Ph.D. student of Arabic literature in Aleppo, but was forced to abandon her studies back in 2015. Her family had paid for her and her husband to travel first to Lebanon, then to Egypt. Egypt became a difficult place for Syrians to live after el-Sisi rose to power, so they flew to Algeria, then crossed illegally into Morocco; a process which caused the couple great suffering ('You have to understand Emma, we've never broken the law before'). Samar took out two books from her bag, which she then handed to me with great pride: collections of modern Arabic poetry which she had edited whilst still in Aleppo. She must have carried several copies of these books all the way from home. On the inside cover, she wrote a message: 'Dearest Emma, beautiful mother and scholar. With love from Samar'. I beamed at receiving the gift, and the compliment.

Sociologising these encounters: How to avoid reproducing the 'good' refugee discourse?

How could I sociologise and objectivise these situations in which I found myself? Of course, there were many ways for the story of 'third country nationals' stranded in Morocco to be told. The incident of the public flogging didn't get written into the final 'deliverable' or journal article. However much I tried to reflect upon this and theorise it, the encounter never managed to fit congruously with the project's aim of re-humanising people who had been de-humanised. There was no space for complexity, uncertainty and even contradiction in my analysis; this dissonance could only ever be a residue which needed to be wiped away (cf. Morin 1992). I was also explicitly warned by colleagues not to go near the topic; 'You don't have a permanent job yet, think how this would come across on social media if someone misinterpreted you in a conference…'. No amount of reflection on power relations and ethnocentric assumptions could mitigate against the good vs. bad migrant narrative which could emerge from juxtaposing encounters in Fez and Rabat.

Later, and after a good few of the late-night drunken chats I alluded to earlier, I came to reflect upon the way the violent practices of the Schengen border were mimicked and reproduced through the violence of the camp; from the categorising and separating of people based on their nationalities, to the beatings and humiliation of those deemed to have transgressed the rules. In a way, the category of the 'black migrant' became a lived-in category and went some way in 'making up' these people on the move (Hacking 2006; Andersson 2016; Tazzioli 2019). The levels of violence that they had faced (one man from Ghana had lost his legs from an attempt to scale the wall at Ceuta) did nothing to reduce the desire to come to Europe. So perhaps in many ways, Sarah was right to chastise me for deciding to focus my research on the Levantine Arabs. An analysis based solely on mimesis however removed something from the story of these men living in the camp. I didn't want to write a romanticised, homogenised story of migrant autonomy or reproduce this valorisation of suffering.

Writing about Syrians who were 'like us', was comparatively far easier (see also Žižek 2016 on this subject). Samar had the symbolic capital to be able to manipulate our encounter somewhat, to ethnographically 'seduce' me (cf. Tubaro and Casilli 2010), writing a personal note to me in her book though she'd only known me for three or four hours. Her education, gender, and confidence meant that she was able to temper and soften the power relations between us. In this encounter, I wasn't the European cosmopolitan from the research business, making a living out of her suffering. At least I wasn't *only* that. I was also a fellow scholar, young woman, and mother. Plus, she was nice to me and complimented my appearance. This was altogether a much less uncomfortable set of relations.

What also didn't get written into the final write-up however was the disdain felt by many of the Syrians towards the black Africans in Morocco. It wasn't only Sarah, but several other NGO workers and volunteers that complained about the rude behaviour of the Arabs towards the other migrants, the complete normality of overt racism and hostility towards Africans, which sometimes escalated to minor scuffles. This prejudice wasn't exactly far hidden in Samar's story, but who was I to judge? She was glamorous and educated. The Syrians were also suffering.

Reflexivity is supposed to be the answer to all of this. And in many ways, a careful, empirically situated analysis of relations is supposed be so much more than self-flagellation or a descent into narcissism (see Hamarti-Ataya 2013 for this discussion). Shedding light on the assumptions, biases and baggage that one brings to the field is an indispensable first step. But with a strong ethnographic analysis, perhaps it's possible to trace the ways in which the research encounter is itself shaped by the researcher's positionality, beyond the seemingly sterile and bureaucratic way of 'objectifying the object' (Bourdieu 1988). I carried with me a multiplicity of subjectivities into the field; a foreign mercenary, an expert, a mother, or a fellow scholar. With an attentive ethnographic ear, one can follow the shifts between these possibilities, as well as the activation of positional relations that hadn't been anticipated. Opening myself and my fieldnotes in this way to others; to productive misunderstandings and alternative interpretations – is as much a part of the anthropological venture as is conducting fieldwork (Fassin 2017). Sometimes I forgot that reflexivity is a collective enterprise and that nobody can do it on their own. The late-night drunken chats are part and parcel of the endeavour.

Conclusion: Rethinking reflexivity in times of precariousness

By way of conclusion, I'd just like to dwell a bit more on what reflexive scholarship could mean in light of how we conceive 'failure'. A practical and collective reflexivity, with the help of friends and colleagues, is indispensable to the type of embedded, ethnographic fieldwork so many of us are now undertaking in the field of IR.

What I'm still unsure about however, is whether this commitment to reflexivity can mitigate against the necessities of the early career scholar in precarious employment to refrain from saying the wrong thing; being too risk averse to take a chance. Admitting when something hasn't gone to plan; that fieldwork can be confusing, frightening, destabilising; that your interlocutors find you repulsive; – these issues shouldn't be the big taboos that they are. In this sense, we are all a little bit complicit in maintaining a certain 'public transcript'; a carefully choreographed performance of what acceptable research should look like, which serves to uphold certain types of domination and exclusion.

I can't imagine soliciting contributions for this volume was especially easy for the editors, but provoking and reinvigorating conversations about failure, however we think about it, far away from the 'think-about-how-this-will-affect-your-career' contingent is a necessary and welcome undertaking. Perhaps acknowledging these taboos, these anxieties and these pressures as precarious scholars could allow us to revisit what we mean by the very idea of reflexive scholarly practice.

References

Andersson, Ruben. 2016. 'Europe's Failed 'Fight' against Irregular Migration: Ethnographic Notes on a Counterproductive Industry'. *Journal of Ethnic and Migration Studies* 42 (7): 1055–75.

———2014. *Illegality, Inc.: Clandestine Migration and the Business of Bordering Europe*. Berkeley: University of California Press.

Appadurai, Arjun. 1997. 'Discussion: Fieldwork in the Era of Globalization'. *Anthropology and Humanism* 22 (1): 115–118.

Basaran, Tugba, and Elspeth Guild. 2016. 'Mobilities, Ruptures, Transitions'. In *International Political Sociology: Transversal Lines*, edited by Tugba Basaran, Didier Bigo, Emmanuel-Pierre Guittet, and R. B. J. Walker, 272–285. London and New York: Routledge.

Bigo, Didier, and Emma Mc Cluskey. 2017. 'What Is a Paris Approach to (in) Securitization? Political Anthropological Research for International Sociology.' In *The Oxford Handbook of International Security*, edited by Alexandra Gheciu and William C. Wohlforth, 116–130. Oxford: Oxford University Press.

Bourdieu, Pierre. 1988. *Homo academicus*. Stanford: Stanford University Press.

Cohen, Stan. 2001. *States of Denial: Knowing About Atrocity and Suffering*. London: Polity.

Eckl, Julian. 2008. 'Responsible Scholarship after Leaving the Veranda: Normative Issues Faced by Field Researchers and Armchair Scientists'. *International Political Sociology* 2 (3): 185–203.

Fassin, Didier. 2017. 'The endurance of critique'. *Anthropological Theory* 17 (1): 4–29.

FRONTEX_European_Border_and_Coastguard_Agency. 2018. 'Migratory Route Map'. https://frontex.europa.eu/along-eu-borders/migratory-map/.

Gabrielsen Jumbert, Maria. 2013. 'Controlling the Mediterranean Space through Surveillance. The Politics and Discourse of Surveillance as an All-Encompassing Solution to EU Maritime Border Management Issues'.

Espace populations sociétés. Space populations societies 2012/3: 35–48.

Geertz, Clifford. 1984. 'Anti anti-relativism'. *American Anthropologist* 86 (2): 263–278.

Hacking, Ian. 2006. 'Making Up People'. *London Review of Books* 28 (16): 23–26.

Hamati-Ataya, Inanna. 2013. 'Reflectivity, reflexivity, reflexivism: IR's "reflexive turn" and beyond'. *European Journal of International Relations* 19 (4): 669–694.

Hanson, Rebecca and Patricia Richards. 2017. 'Sexual Harassment and the Construction of Ethnographic Knowledge'. *Sociological Forum* 32 (3): 587–609.

Heck, Gerda, and Sabine Hess. 2017. 'Tracing the Effects of the EU-Turkey Deal'. *Movements. Journal for Critical Migration and Border Regime Studies* 3 (2): 35–56.

Jeandesboz, Julien, and Polly Pallister-Wilkins. 2016. 'Crisis, Routine, Consolidation: The Politics of the Mediterranean Migration Crisis'. *Mediterranean Politics* 21 (2): 316–20.

Madison, D. Soyini. 2011. *Critical ethnography: Method, ethics, and performance*. London: SAGE publications.

Montesinos Coleman, Lara. 2015. 'Ethnography, Commitment, and Critique: Departing from Activist Scholarship'. *International Political Sociology* 9 (3): 263–80.

Morin, Edgar. 1992. 'From the concept of system to the paradigm of complexity'. *Journal of social and evolutionary systems* 15 (4): 371–385.

Olwig, Karen. 2007. *Caribbean Journeys: An Ethnography of Migration and Home in Three Family Networks*. Durham: Duke University Press.

Pallister-Wilkins, Polly. 2015. 'The Humanitarian Politics of European Border Policing: Frontex and Border Police in Evros'. *International Political Sociology* 9 (1): 53–69.

Picozza, Fiorenza. 2018. 'Tracing Europe's Geographies of Asylum: Coloniality, Mobility and Solidarity During and after the 2015 "Refugee Crisis"'. King's College London.

Tazzioli, Martina. 2019. *The Making of Migration: The Biopolitics of Migration at Europe's Borders*. London: SAGE Publications.

Tubaro, Paola, and Antonio A Casilli. 2010. '"An Ethnographic Seduction": How Qualitative Research and Agent-Based Models Can Benefit Each Other'. *Bulletin of Sociological Methodology/Bulletin de Méthodologie Sociologique* 106 (1): 59–74.

Žižek, Slavoj. 2016. 'The Cologne Attacks Were an Obscene Version of Carnival'. *New Statesman* 13.

8

Failing in the Reflexive and Collaborative Turns: Empire, Colonialism, Gender and the Impossibilities of North-South Collaborations

DESIRÉE POETS

I am a half-German, white, cis settler who was born and raised in Brazil and was working with urban indigenous and black communities there. At the same time, I was institutionally located as a Ph.D. candidate in Wales, in the UK. In such a project, questions of ethics and accountability were imperative. As I describe below, these imperatives quickly led me to self-reflexivity, the practice of being transparent and reflexive about one's positionality in the field and how it affects the research design and process. They also led me to collaborative methods, which, in the words of Himika Bhattacharya (2008, 305), are about 'doing ethnography "with" people rather than "on" or "about" people, with a purpose of bringing about positive change in the lives of the researched'. Despite the important contributions that such methodologies make to ethical and accountable research, I incurred certain failures in this process that were both my own and part of academia more broadly. I will focus here on two such failures. The first was my falling (or failing) into what D'Arcangelis (2017) called 'the White settler fantasy of transcending colonialism'. The second was my failure to openly reflect on sexualised encounters and sexual harassment in the field.

The story I tell here, although it is built somewhat chronologically, should not be read as a linear progression towards 'better' understanding and 'better' research practices. The fieldwork and my Ph.D. as a whole were an uneven process of theoretical and empirical trial and error with no guarantees. They included experiences that I was only able to articulate and name after the fact, and difficulties or impossibilities of which I may have been intellectually aware, but only comprehended on a more affective level after experiencing them myself.

Two occasions made my failures evident. The first was during one of my returns to Brazil to share my analysis with the groups with whom I had worked. During one such meeting, a female member of a *quilombo*, a history teacher and postgraduate student, rolled her eyes at me and challenged the overly academicist focus of my presentation. What she wanted instead were political strategies that would support the community. In response, I put my notes down and quickly rethought my experiences in a different light to answer her question. But, I thought, why were *those* findings not already central in my Ph.D. dissertation? To what extent was the Ph.D. itself ultimately useful to the groups? The second occasion was when my then Head of Department contacted me about a prize for 'research impact' and asked whether I could demonstrate any direct impact of my work and apply for the prize. Leaving aside the problems with the 'research impact' agenda in the United Kingdom, the truth was that I could not demonstrate any. The only tangible outcome of my collaborations had been my Ph.D. title and, more recently, an early academic career in the US. The struggles and realities of the communities in Brazil remained more or less unchanged by my project. This is a very real material outcome of my collaborations, which repeatedly put me face to face with the extractivism of my work – an extractivism which collaborative and reflexive research methods between North and South run the risk of renewing and re-legitimising.

The first limit to my collaborations was my research design. For example, I had applied to the Ph.D. programme with a project I had designed on my own. Working between three cities and with four groups, I had limited time to collaborate with each. And my central research question, the anchor of any International Relations dissertation, namely 'Who counts as indigenous?', was a white, settler question. As to my second failure, besides my research design, gendered and sexualised encounters also shaped and limited my collaborations. Sexual harassment, for example, stopped me from engaging more fully with one of the groups. At the time, I tried to ignore these encounters, not mentioning them to my supervisors or colleagues, and left them out of my dissertation. This is a common response of female researchers who experience harassment in the field. Returning to these experiences for this chapter, I therefore understand these failures as my own, but also as part of the racialised, heteropatriarchal, and colonial foundations of the academic industrial complex (Stein 2018; Grosfoguel 2013).

Despite decades of critical and feminist critiques, these foundations continue to invisibilise sex, gender, and the body in knowledge production (Hanson and Richards 2019), normalising and/or obfuscating sexual harassment both in the academy and in the field. They can also co-opt reflexive and collaborative research methods to reproduce racial/colonial hierarchies in seemingly more benign forms, replacing structural transformations in the academy and the society in which it is located (here) with collaboration, solidarity, reflexivity, and activism in the field (out there) only. Yet, the field and the academy are continuations of one another, so that our failures here are our failures there. I am not alone in this discomfort and the dilemma that such North-South engagements bring in the era of the decolonial, post-colonial, reflexive and collaborative turns. Sam Halvorsen (2018, 12) describes this discomfort as an 'ethicopolitical moment' that has led many, myself included, 'to revisit the otherwise implicit value of learning from [and, in my case, collaborating with] the South'.

Coming to the reflexive and the collaborative turn

My undergraduate degree in International Relations did not include methods training. When I came to research methods and design courses during my Master's, these were either taught from an overtly theoretical, critical realist perspective, or were University-wide courses that did not go in depth into specific methods. I came from a theory-heavy department in a discipline that already did not prioritise ethnographic fieldwork with marginalised groups, and in which indigeneity and race were – and still are – invisibilised. Although I had engaged with indigenous and feminist methodologies through, for example, Linda T. Smith (1999) and Gayatri C. Spivak (1988), I only came to feminist ethnography after the end of my first year as a Ph.D. student, when I had already designed my project and fieldwork.

Feminist ethnography is still one of the most productive sights of fieldwork training and theorising for those of us engaged in research with minoritised groups. Emerging in the 1970s and 80s, it has been concerned, alongside matters of gender and sexuality, with questions of voice, representation, objectivity, and power in the research process. A diverse field, it unpacks how research is always marked by the researcher's positionality (the intersections of race, gender, class, religion, sexuality, institutional location, and so on) in relationship with the researched (Haraway 1988; Visweswaran 1997). Feminist ethnography centres 'the basic political issue at the heart of most anthropology – the issue of Western knowers and representers, and non-Western knowns and represented' (Abu-Lughod 1990, 11). Part of the West/non-West dichotomies, as Abu-Lughod (1990) argues, are issues of self/other that contain hierarchies of power that force us to pay attention to the political implications of knowledge production.

One of the most helpful lessons of feminist ethnography is its attention to self-reflexivity, the practice of being transparent and reflexive about one's positionality in the field and how it affects the research design and process. While self-reflexivity was already part of my research, feminist ethnography helped me deepen it in practice. Throughout my fieldwork and writing process, I engaged with the task of reflecting on how the intersecting differences of class, colour, institutional location, gender, and others were shaping my experiences in the field – although, as I discuss later, I excluded any reference to sexual harassment in my final drafts. I was, however, also mindful of how Western academia had been historically instrumental to colonialism and Empire, fixing the non-European world as its object of study while extracting knowledge from it. Besides being reflexive about this reality, I wanted to avoid reproducing these dynamics in my own 'home' country.

To this end, I contacted all groups with the explicit offer 'to make myself useful'. This offer was gradually taken up: I organised a community's archives, helped with a community's online presence, and fulfilled other small tasks I was asked to do. I also began to write for an NGO-run activist journalism platform, often in defence of the communities with whom I was working. It was through this activist research practice during fieldwork that I came to the literature on the collaborative turn (Bhattacharya 2008; Lassiter 2005) at the end of my second year, when trying to make sense of the solidarities I was building on the ground. That was the summer of my final longer fieldwork trip, after which I was supposed to start the write-up phase and final year of my Ph.D. That literature pushed me a bit further and, during that summer, I agreed with all the groups that I would return before and after my Ph.D. defence to present my findings, allow for their interventions, and to report on the defence.

Feminist ethnography, such as the work of Richa Nagar (2014), also reminds us that the inequalities implied in the intersecting differences that make up our positionalities cannot be done away with in the field, including in collaborative and solidarity work. Such work is therefore marked by inevitable impossibilities and difficulties, and the ever-present risk of epistemic violence. I was aware of these conditions of possibility, and yet – as will become clear in the next section – I was not prepared for their effect on my scholarship. During some of the 'return' meetings it became clear that the limits of any collaboration in the field were already set by the very design of my Ph.D., including the central research question, and that the actual contents of my dissertation were only marginally interesting and not immediately useful to the communities.

On not dismantling the white, settler house

Sometimes when I tell this story, my interlocutors try to rescue me by pointing out how much I have done to address the issues I raise here and/or read it as an expression of individualising and even navel-gazing white guilt. Yet what I found myself embedded in were, quite to the contrary, power structures that mould academic knowledge production and in which self-reflexivity and collaborative methods are entangled. These structural entanglements persist despite my individual efforts to address them, as I discuss below. There was a moment in my fieldwork when my embeddedness in these structures was made clear to me, and which could have, or should have, given me pause. When I was contacting groups in Rio during my first year, I spoke to a maroon leader who started our conversation by asking numerous questions about my race, religion, foreign language skills (beyond Portuguese), and my ability and availability to help him with projects for the community. After my answers, he quickly said 'no'. I respected that 'interview process' and his decision, but, in my anxiety to complete the Ph.D. requirements, I also tried to 'move on' quickly, focusing on not thinking about his refusal too much. Yet, this failure to gain access was important in ways that I only later began to comprehend.

Today, I understand his refusal as telling me that my 'good intentions' and 'perseverance' could not undo my structural position and the asymmetry of our potential collaboration, one which he was not interested in reinforcing or validating. Although this thought did cross my mind when 'access' was denied to me, I did not linger with the effects of my inevitable complicity. Part of me still, perhaps subconsciously, believed that through self-reflexivity, collaborative methods, and solidarity-building I could make my research more benign, less problematic, or somehow not as implicated in racial/colonial structures – if only I were given the chance (see D'Arcangelis 2017, 350). I fell into the trap of what Carol Lynne D'Arcangelis (2017), in her own critique of self-reflexivity, called 'the White settler fantasy of transcending colonialism' (340). This fantasy is tied to a desire for innocence which, as she puts it, is anchored in 'modernist/liberal imaginings of a subject capable of transcending structural power inequalities' (342). Even though I was intellectually aware of the arguments otherwise, I had not lingered with the *affect* of that complicity until I reached the write-up phase, when I had to retreat from more activist initiatives and systematised my work in a way that felt entirely out of context.

My project, too, was extracting value from the Global South and my own 'home' country. This came in the form of both symbolic capital (a doctor title) and material capital through my Ph.D. funding and, later, an academic position in the US. I would even argue that it was precisely my collaborative and feminist methods in the Global South that set me apart in the competitive academic job market – an uncomfortable illustration of the structural contradictions of my work. I found myself removing the communities' local knowledges from their immediate political contexts – which Sam Halvorsen (2018) identified as their use-value for grassroots movements – to introduce them into Anglophone academic circuits, where they accumulate exchange value in a global academic labour division that privileges scholars in the Anglophone world (Halvorsen 2018). It was unclear what exactly the groups or even Brazilian academia were gaining from this, despite my 'local' activism and solidarity. In the incisive critique of Judith Stacey (1988, 23), 'the lives, loves, and tragedies that fieldwork informants share with a researcher are ultimately data, grist for the ethnographic mill, a mill that has a truly grinding power'.

Moreover, and beyond my individual career and material gains, how were my good intentions in the field actually re-legitimising the Western academy more generally, making it only seemingly more benign? To paraphrase Tuck and Yang (2014), how were my collaborative methods and self-reflexivity masking a political economy that reproduces racial/colonial power relations to allow the Western academy to accumulate more and more territory in the Global South? There was no self-reflecting and collaborating out of these dynamics. Having said this, my admission here does not undo my complicity either and deeper self-reflexivity will not resolve it. As Sara Ahmed (2004, point 4; emphasis in original) put it, 'the work of critique does not mean the transcendence of the object of our critique; indeed, *critique might even be dependent on non-transcendence*'. Keeping with this theme, Andrea Smith (2013, 266–267) also reminds us that such moments of confession can constitute rather than challenge the settler/white subject through the 'raw material of the Native':

> A typical instance of this will involve non-Native peoples who make presentations based on what they 'learned' while doing solidarity work with Native peoples in their field research/solidarity work. Complete with videos and slide shows, the presenters will express the privilege with which they struggled. We will learn how they tried to address the power imbalances between them and the peoples with which they studied or worked. We will learn how they struggled to gain their trust. Invariably, the narrative begins with the presenters initially facing the distrust of the Natives because of their settler/white privilege. But through perseverance and good intentions, the researchers overcome this distrust and earn the friendship of their ethnographic objects.

In this act of self-reflexivity, the white/settler subject comes into being through how they affect the Other – the white settler's self-reflexivity is made possible through the Other (D'Arcangelis 2017; Wasserfall 1993). This is an ever-present risk in my work, and it is not to dismiss self-reflexivity or *individual* acts to address asymmetries. It is to recognise that, as individual acts within wider structures, they remain complicit. It is not to replace individualised with structural actions either, in a move that dismisses the role of the individual scholar in reproducing systemic conditions. It is to see the individual researcher and the structure as entangled, and therefore to see transformations on one level only as limited without transformations on the other.

Gendered and sexualised failures

Empire and settler colonialism are also co-constituted through heteropatriarchal gender relations (Simpson 2016; Simpson 2017; McClintock 1995), which equally shape the academy. There (or here), assumptions of the 'standard' researcher or scholar as still white, cis, male, and usually from the West have also informed how ethnographic methods have been developed, despite decades of critical and feminist methods, as Richards and Hanson (2019, 25) have argued in their timely qualitative study with women fieldworkers. They show how these assumptions have given rise to standards for what counts as 'good ethnography' – ethnography that still often invisibilises gender, sex, and the body in the field, especially for researchers of colour and LGBTQIA+ researchers. This is why, they argue, discussions of sexual harassment in the field remain circumscribed to feminist circles, leaving women who experience harassment with an 'awkward surplus', stories 'which can be both difficult and risky to fit into our findings and theories [and that] become superfluous stories, excess that must be cut to get to the 'real' data' (Richards and Hanson 2019, 2–3).

These standards take the form of 'ethnographic fixations' that can place researchers in harm's way. Such fixations include solitude, the assumption that ethnography is an individual endeavour; danger, which includes the ongoing glorification of dangerous ethnographies; and intimacy, the idea that 'good ethnographies' are ones that stem from intimate relationships between researcher and researched (Richards and Hanson 2019, 28–39; see also Introduction, this volume). During my time as a Ph.D. student, I would pendulate between wishing I could replicate the fieldwork experiences of my male colleagues of bonding with (often other male) 'informants' in the field, which seemed to regularly take the form of excessive drinking, and those of female researchers of building relationships of trust, solidarity, and intimacy with (often female) individuals or groups. Cases where a researcher may not be in the place to build such intimacies or where her safety is not assumed seemed limited to research with extremist (e.g. right-wing) groups or situations of war and conflict (see Sriram et al. 2009). Discussions of such dynamics within progressive movements and in 'everyday life' are still relatively uncommon, although as soon as the topic emerges in a group of female researchers, the stories begin to flow.

As is still regularly the case, leaderships in the social movements with whom I was working were mostly male. In one case, they were all male. Looking back, I suspect that I was granted 'access' to this group, in part at least, as a result of my being a young, female student, and therefore not taken seriously or read as in any way 'threatening'. Gendered, racialised, and sexualised relations can therefore both enable and hinder one's research and have divergent effects on the researcher's 'access' and safety, depending on their positionality. For example, there were also situations in the field where I felt safer for being a white, foreign-looking woman, such as during protests. There was one case, however, in which my research with a social movement was marked by a series of uncomfortable comments, encounters, and interviews that led me to retreat from deeper engagement. One such situation was when I was invited to see a movie in which many of the group members had acted with the explicit confirmation that it would be a group outing, only to find myself on a one-on-one 'date'. Another was when I accepted a lift after an informant had interrupted an interview suggesting we 'go to lunch instead', during which he asked repeated questions about my love life – which is not uncommon for female researchers (see Freitas et al. 2017). I ended up having to jump out of the car as soon as we came to a red light because, without explanation, he simply passed where I was supposed to be dropped off.

After these incidents, I avoided one-on-one interactions and agreed to attend only public and collective meetings, thereby closing off possibilities for solitude, danger, and intimacy. Yet, I did not directly confront the men involved in these incidents, trying not to be 'unpleasant', and allowing several subsequent degrading or sexist comments about me to pass. I was scared that reacting to these may jeopardise my access. My collaboration with them was therefore more limited and ended up taking the form of newspaper articles in support of their movement. I also did not include a discussion of these incidents in the final draft of my Ph.D., failing to reflect on how these interactions shaped my findings. Part of the reason for this was that it seemed inappropriate to disclose them in a dissertation, also for their potential to reinforce racialised stereotypes and to centre me and my whiteness in the project as a victim. While occupying a relatively vulnerable position as a woman, I was also in a position of privilege in terms of my class, race, and institutional location. These shifting power relations could not be easily resolved, so I chose to not discuss these incidents at all.

Another reason for not discussing them was that I felt that, because I 'knew' Brazil and was 'used to' the sexism 'there', these encounters were unremarkable. Why would this 'everyday sexism' be worth writing about? I could almost hear my mother, herself with many accounts of sexual harassment 'at home', saying to me: *Well, what did you think was going to happen?* It was just part of life. Yet, this 'part of life' is missing from usual accounts and training guidelines on collaborative and activist methodologies, which obfuscate more complicated and difficult solidarities. Although gendered and sexualised dynamics in the field are being increasingly written about (Bell, Caplan, and Karim 1993; Freitas et al. 2017), also with attention to race (Berry et al. 2017; Hanson and Richards 2019), the literature on ethnography still assumes either a 'racially privileged male anthropologist' (Berry et al. 2017, 538) or a relationship of trust and safety between researcher and researched that enables 'radical vulnerability' in solidarity-building (Nagar 2014).

The obfuscation of sexual harassment in the field, as Hanson and Richards (2019) observe, is inseparable from academia's own patriarchal and sexist structures, which dismiss sexual harassment allegations in the still male-dominated academy itself – something ongoing denouncements in several academic circles have shown (see Ahmed 2017). Sexual harassment in the field hereby becomes 'a "given," just one more hardship worth navigating to gather good data' (Hanson and Richards 2019, 2). My failure to address harassment in the field (*there*) is therefore also tied to academia's own power relations (*here*). Nonetheless, Hanson and Richards add that the lack of training on gendered fieldwork dynamics is part of the 'colonialist legacy of ethnography', with its roots in colonial and Imperial expansion. This is expressed in 'the assumption that researchers can somehow stand above and beyond the community they study' (21), in a renewal of expectations of disembodied, observant neutrality. The writing out of such gendered and sexualised encounters in the field is both a product of and a means of reproducing this colonial illusion of neutrality that persists despite the turns to self-reflexivity and feminist as well as critical methodologies – which, generally, continue to be marginalised in most fields, including (or perhaps especially) in International Relations.

Conclusion

In his critique of recent calls and movements to 'decolonise the academy', Andile Mngxitama (2018) asks, 'is a decolonised University possible in a colonial society?' His answer is 'no'. Efforts to transform the academy, he states, must be tied to efforts to transform society. Similarly, neither are decolonised research methods possible in a colonial society and University. Our collaborative and activist work in the field (*out there*) equally fails if it is not continued in efforts to address the colonial, racial, and heteropatriarchal designs of the academic industrial complex and academic knowledge production (*here*). This includes transformations in and beyond the classroom that challenge the ongoing lack of representation as well as the dominant norms and standards for what counts as 'good ethnography' that obfuscates how ethnography is inevitably embodied and, therefore, racialised and gendered. Within this, they also reproduce the historical invisibilisation and marginalisation of researchers who do not conform to standards of whiteness, heteronormativity, and masculinity.

Relatedly, as I enter my early career and continue to work through what it means to do politically meaningful research, I have come to the following guiding question: As Postcolonial and Decolonial Theory as well as feminist and collaborative methods seem to gradually become accepted in the mainstream, how will we make sure that these movements are not reduced to means through which academia aims to re-invent itself as only seemingly more benign? In responding to such moves to decolonise the academy, the Brazil-based indigenous thinker Ailton Krenak (2019) described them as a haemodialysis, in which colonial academic institutions increasingly run by market logics 'take someone else's blood to keep on working' while we all remain 'immersed in Coloniality' as environmental and political crises continue to unfold. This is the ever-present risk that my collaborative and self-reflexive methodology must navigate.

My response to this dilemma has been to turn my efforts to collective action across differences (see hooks 1986) in my current region and institution, alongside my ongoing activism in Brazil, to gradually contribute to the structural change necessary for my methods to be effective. There is a long road ahead, but guiding me, here, are the words of Andrea Smith (2013, 264):

> ...individual transformation must occur concurrently with social and political transformation. That is, the undoing of privilege occurs not by individuals confessing their privileges or trying to think themselves into a new subject position, but through the creation of collective structures that dismantle the systems that enable these privileges.

I finish here with six acknowledgements to my younger self that may be helpful to fellow fieldworkers:

1. No matter your positionality, your fieldwork is embodied. Pay attention to how your body is shaping your fieldwork: What spaces does it open up for you, and how, and which ones does it close off? Whom are you drawn to, and why? Who is drawn to you, and why?
2. Put your safety first. If you have a gut feeling that something is amiss, you are probably right. Reflect on when the 'ethnographic fixations' of danger, solitude, and intimacy, which Richardson and Hanson (2019) discuss, are potentially putting you in harm's way.
3. The academy, the field, and the societies in which both are nested are continuations of one another, not boundaries.
4. This means that your scholarship is entangled in wider structures of the academic and non-academic world(s). No matter how much you try to address these entanglements, you cannot undo them on your own. We don't come to the field innocently. Think about what this means for your scholarship but also what this means for structural and collective transformation that cannot be reduced to your scholarship or one site only.
5. Let the scholarship and leadership of historically marginalised thinkers and fieldworkers guide you. As the most affected by hegemonic norms and standards, they are the experts and will shine light on where to go. This includes but goes beyond a politics of citation.
6. Finally, you are not alone in the challenges, dilemmas, and failures of fieldwork. Find and build those solidarities, for they will carry you forward. It's a path full of contradictions, complicities, and mistakes – but it doesn't have to be a lonely path.

References

Abu-Lughod, Lila. 1990. 'Can There Be A Feminist Ethnography?' *Women & Performance: A Journal of Feminist Theory* 5 (1): 7–27.

Ahmed, Sara. 2017. *Living a Feminist Life*. Durham: Duke University Press.

———. 2004. 'Declarations of Whiteness: The Non-Performativity of Anti-Racism'. *Borderlands* 3 (2). http://www.borderlands.net.au/vol3no2_2004/ahmed_declarations.htm.

Bell, Diane, Pat Caplan, and Wazir Jahan Karim, eds. 1993. *Gendered Fields: Women, Men, and Ethnography*. New York: Routledge.

Berry, Maya J., Claudia Chávez Argüelles, Shanya Cordis, Sarah Ihmoud, and Elizabeth Velásquez Estrada. 2017. 'Toward a Fugitive Anthropology: Gender, Race, and Violence in the Field'. *Cultural Anthropology* 32(4): 537–565.

Bhattacharya, Himika. 2008. 'New Critical Collaborative Ethnography'. In *Handbook of Emergent Methods*, edited by Sharlene Nagy Hesse-Biber and Patricia Leavy, 303–324. New York and London: Guilford Press.

D'Arcangelis, Carol Lynne (2017). 'Revelations of a White Settler Woman Scholar-Activist: The Fraught Promise of Self-Reflexivity'. *Cultural Studies ↔ Critical Methodologies* 18 (5): 339–353.

Freitas, Caroline Cotta de Mello, Rafaela Nunes Pannain, Heloisa Marques Gimenez, Sue A. S. Iamamoto, and Aiko Ikemura Amaral. 2017. 'Campo, gênero e academia: Notas sobre a experiência de cinco mulheres brasileiras na Bolívia'. *Cadernos de campo* 26 (1): 348–369.

Grosfoguel, Ramón. 2013. 'The Structure of Knowledge in Westernized Universities: Epistemic Racism/Sexism and the Four Genocides/ Epistemicides of the Long 16th Century'. *Human Architecture: Journal of the Sociology of Self-Knowledge* 11 (1): 73–90.

Halvorsen, Sam. 2018. 'Cartographies of epistemic expropriation: Critical reflections on learning from the south'. *Geoforum* 95: 11–20.

Hanson, Rebecca, and Patricia Richards. 2019. *Harassed: Gender, Bodies, and Ethnographic Research*. Oakland, CA: University of California Press.

Haraway, Donna. 1988. 'Situated Knowledges: The Science Question in Feminism and the Privilege of Partial Perspective'. *Feminist Studies* 14 (3): 575–599.

hooks, bell. 1986. 'Sisterhood: Political Solidarity between Women'. *Feminist Review* 23 (1): 125–138.

Lassiter, Luke Eric. 2005. *The Chicago Guide to Collaborative Ethnography*. Chicago: University of Chicago Press.

McClintock, Anne. 1995. *Imperial Leather: Race, Gender and Sexuality in the Colonial Conquest*. New York: Routledge.

Mngxitama, Andile. 2018. 'Is a Decolonised University Possible in a Colonial Society?' In *Routledge Handbook of Postcolonial Politics*, edited by Olivia U. Rutazibwa and Robbie Shilliam, 335–349. London: Routledge.

Nagar, Richa. 2014. *Muddying the Waters: Coauthoring Feminisms Across Scholarship and Activism*. Urbana: University of Illinois Press.

Simpson, Audra. 2016. 'The State is a Man: Theresa Spence, Loretta Saunders and the Gender of Settler Sovereignty'. *Theory & Event* 19 (4).

Simpson, Leanne Betasamosake. 2017. *As we have always done*. Minneapolis: University of Minnesota.

Smith, Andrea. 2013. 'Unsettling the privilege of self-reflexivity'. In *Geographies of Privilege*, edited by France Winddance Twine and Bradley Gardener, 263–279. New York and London: Routledge.

Smith, Linda Tuhiwai. 1999. *Decolonizing methodologies: Research and indigenous peoples*. London and New York: Zed Books.

Spivak, Gayatri Chakravorty. 1988. *Can the subaltern speak?* Basingstoke: Macmillan.

Sriram, Chandra Lekha, John C. King, Julie A. Mertus, Olga Martin-Ortega, and Johanna Herman, eds. 2009. *Surviving Field Research: Working in Violent and Difficult Situations*. London: Routledge.

Stacey, Judith. 1988. 'Can there be a feminist ethnography?' *Women's Studies International Forum* 11 (1): 21–27.

Stein, Sharon. 2018. 'Confronting the Racial-Colonial Foundations of US Higher Education'. *Journal for the Study of Postsecondary and Tertiary Education* 3: 77–96.

Tuck, Eve, and K. Wayne Yang. 2014. 'Unbecoming Claims: Pedagogies of Refusal in Qualitative Research'. *Qualitative Inquiry* 20 (6): 811–818.

Visweswaran, Kamala. 1997. 'Histories of Feminist Ethnography'. *Annual Review of Anthropology* 26: 591–621.

Wasserfall, Rahel. 1993. 'Reflexivity, feminism, difference'. *Qualitative Sociology* 16 (1): 23–41.

9

Reproducing the European Gaze through Reflexivity: The Limits of Calling Out Failures

EWA MACZYNSKA

This text problematises the notion of 'failure', articulated as a condition of critique, through its reliance on and reproduction of reflexivity as a positioned practice advocated by critical theory scholars (Rose 1997; Halberstam 2011; Sjoberg 2018). It does so by analysing a fieldwork experience that I have trouble making sense of because of the way in which it posed a challenge to my practice of reflexivity. With this experience my project faced what Visweswaran, following Spivak, calls 'its own impossibility' in that I was not able to make sense of where I failed without running the risk of reproducing the very dynamics that led me to fail in the first place (Visweswaran 1994, 99). The text starts with the recollection of a fieldwork encounter that I had trouble making sense of and goes on to think through the consequences this encounter might have for problematising reflexivity and failure in fieldwork conducted among marginalised populations.

Reflexivity and marginalisation

It was the summer of 2015 and I was wandering around Copenhagen with Hassan, a 20-year-old guy from Syria holding refugee status in Denmark.[1] We met during one of my first days in Denmark through a Copenhagen-based organisation supporting LGBTQ asylum seekers, and soon started spending most of our time together. The organisation served as an entry point to my field research, as I was interested in analysing narratives of queer asylum seekers in Denmark – people whose voices I saw as marginalised/silenced in

[1] In order to ensure the anonymity of my respondent, I am using the name 'Hassan' instead of his actual name.

hegemonic discourses around migration. I was preparing to conduct research with a population that was not only facing multiple forms of marginalisation but also a population whose stories, behaviour and representation were subject to repeated evaluations on the side of the state with an intention to (de)legitimise their claims for asylum. I was anxious that any narration I would produce about my respondents could contribute to the racism, xenophobia, homophobia, and marginalisation they were facing, as well as potentially be used by the state to delegitimise their claims for asylum (Patai 1991, 139). Thus, I was aware of the position of power I occupied as a researcher and of my ethical responsibilities in writing about and representing others. I was also anxious about the ways my positionality was reflected in both the knowledge I produced and in the very process of conducting fieldwork.

In navigating these anxieties, reflexivity seemed crucial. I saw reflexivity as a positioned practice that helps us face multiple lines of difference and power hierarchies that the researcher is implicated in, as well as recognise and think through failures, shortcomings, and limitations of the research process. Following Pillow, I was committed to practice a 'reflexivity of discomfort' that is neither comfortable nor focused on success, but rather encourages the researcher to stay 'accountable to people's struggles for self-representation and self-determination' (Visweswaran 1994, 32; Pillow 2003, 193). The act of recognising and embracing failures, not in order to turn them into successes, but rather to see them as spaces for thinking about the research process and developing further reflexivity seemed to be an inevitable element of my work. It was precisely from this position of awareness and sensitivity that I was approaching Hassan.

Hassan

I couldn't be happier to meet Hassan so early into my fieldwork. He was eager to spend time with me, help me organise my interviews, accompany me to asylum centres, and help me understand the workings of the Danish asylum procedure. He was articulate and reflexive about his experiences, and we would often enter long and at times heated discussions mostly about politics and life in Europe. The first few times we met, we treated our meetings as a part of my research. It quickly turned out that we also just got along really well, and soon started spending most of our days together. We became each other's companion. I approached him with liking, compassion and interest, both as a friend and as a significant person for my research.

Over the course of our interactions I learned that he grew up in an upper-class Syrian family in an affluential neighbourhood of Damascus. At the age of 19, he immigrated to Europe through the East African migratory route. His parents were still in Syria. He spoke little about the war but, from what he shared, I understood that he had lost many close people to the war, including loved ones. As a middle class Polish queer woman in her late 20s, my abilities to imagine what it meant to grow up in an affluential neighbourhood of Damascus as a (queer) male member of a very prominent family were limited. Unfamiliar with the context within which he was raised, I was unable to imagine how the complex layers of privilege and oppression, resulting from his class position, gender, religion, and sexual orientation (to name a few), entangled. I was unable to relate to and understand the socio-political structures he grew up in – not only because they were different from mine, but also because in the Eurocentric scholarly context in which I grew up, we were not taught about and trained to understand the contexts of non-Western societies.

I could orient myself, of course, with the complexities of Hassan's socio-political background, but he was one of several other migrants that I was working with, all with different personal and national histories coming from countries like Ghana, Syria, and Afghanistan. Their experiences of migration – that I at first regarded as an important denominator of their stories – quickly proved to be of little help in examining the complex nature of their experiences in Europe and around asylum/migration. In that, I faced a common problem in migration scholarship: migrants' stories always seem to start only at their meeting with the border regimes of Western countries. Apart from not having a first-hand experience of the complex structures of privilege that Hassan grew up in, I was also not able to understand how it felt to have one's life severely impacted by war or to be an Arab migrant in a country with strong Islamophobic discourses. Whether I liked it or not, I understood Hassan's behaviour and words from the narrow scope of interpretation that was available to me as a European researcher.

'I don't like Arabs'

As we were spending more time together, I began realising that I often read Hassan's comments as having racist and/or Islamophobic tones. 'I don't like Arabs' – he would say openly and repeatedly. 'Black people stink' – he commented sometimes. I wasn't able to make sense of the racist tone of the comments: was he trying to provoke other members of the organisation? Was he being sarcastic? Did he mean to be insulting? At first, from Hassan's description of the difficulties he faced while befriending Danish people, I thought that one of the reasons he felt comfortable with me was because I was not Danish and thus similarly an outcast, socially and economically, in the Danish society. As I kept hearing his racist comments, I also began entertaining the possibility that he felt at ease with me not only because I was enough of an outsider in Denmark, but also because I wasn't too much of an outsider. As a white person from a EU country, I did not fit the stereotypical and highly problematic image of the 'migrant' painted within the Danish nationalist discourse – one that Hassan was positioning himself in relation to. Thus, spending time with me wasn't perpetuating his stigmatisation as much as hanging out with Muslims or people of colour would. I was unsure I was able to understand and interpret him and our relationship correctly.

I wasn't sure how to position myself in relation to the racist and Islamophobic comments Hassan made and how to show my disagreement with him without sounding patronising. Trained to think through my positionality, I had to time and again remind myself that any response I give is that of a white European academic to a person of colour with a migration background. Within such a framework, condemning his racism felt as problematic as ignoring it. I was to either try to 'explain' to him why he should not be racist, an approach itself quite racist in its underpinnings, or remain silent in the face of his comments which I worried would make me complicit in a racist narrative. I was quietly negotiating my position between two unappealing options, trying to find responses that would be least likely to contribute to further marginalisation of both Hassan and the people who could be influenced by his offensive comments. I avoided direct confrontation and gave him moderate responses until one particular evening.

Frustration

We were walking back home from a meeting organised by a small support group for LGBTQ people of colour living in Denmark and we were both agitated. I had observed Hassan ostentatiously rolling his eyes, puffing, and making malicious comments towards other participants throughout the duration of the meeting. On the way home, his comments became openly racist and he went on to occupy a position of complete isolation and not belonging. 'I hate those people' – he was almost shouting – 'they are so stupid and so pathetic. I hate those Arabs, have you seen how they look? And they can't even speak proper Arabic. And I know they look down on Syrians. And at this meeting, have you seen how they were looking at the black guy? Everyone in Denmark loves black men. I don't like black people, but in Denmark I wish I was black'. Although it was not the first time I heard him pass unsettling comments, about himself and/or other people – a way of acting out, I assumed – this was the first time he went so far in expressing his anger and the first time I was not able to control my frustration. 'Why do you say all these things? And why did you behave this way?' – I asked angrily, disturbed both by how offensive his comments were and my own inability to respond to them. I was agitated and impatient. Maybe that is exactly what he wanted me to be. 'This is how I feel!' He became even more furious. 'These people are stupid, and they are pathetic. I hate that I need to pretend to be nice, to be someone that I am not, to constantly watch my words. And I hate how Danish people are nice to all the migrants, even though I am sure they hate them, but they pretend to be nice because they need to be politically correct. I hate this political correctness; I hate being nice, pretending to think something I don't. I am not going to be nice to people only because I am not supposed to be racist!'

He went on ranting. His outburst infused in me a sense of confusion – I experienced anger and disappointment mixed with compassion and I did not know how to react. I heard in his words disenchantment, frustration, and isolation. I imagined he spoke out the trauma of being brutally displaced and feeling isolated in a country where his body was constantly subjected to racialisation. But this did not justify or neutralise his racism. I was hearing a young man who suffered displacement but also one who grew up in the upper class of his society, with all the implications that came with such positioning. I heard entitlement, racism, and Islamophobia. I heard him navigate complex hierarchies that he was both suffering from and reproducing. I heard him position himself vis-à-vis various groups; Danish people, migrants, Arabs, the black community, Syrians. I wanted him to feel better, but I also felt discomforted by him working through his frustration at the expense of others. I heard his anger, his disappointment, and what I imagined was a loss of one's life as we know it. With this, I arrived at a point where I posed a key question to myself; why did I first think of him as being displaced and not as being a man who grew up in the upper classes?

Eurocentric framing of identity

I realised that my interpretation of Hassan's behaviour relied on identity markers that became a source of marginalisation in Europe, and not those that structured his life in Syria. I was framing his identity from a Eurocentric perspective; as though it was the experience of being a migrant in Europe rather than the complex history of his life in Syria that framed his political stance and influenced his ways of expressing himself. I couldn't quite locate within this Eurocentric perspective the multiple axes of privilege and oppression that our encounter (and my interpretation of it) and his encounter with other groups (and his and my interpretation of it) were structured around. Was I simply projecting on him my own imaginaries of what it meant to be a migrant? And on myself my own imaginaries of what kind of researcher I wanted to be? As Trinh Minh-Ha (1989, 76) writes, reflexivity defines both the subject written and the subject writing. Weren't my thinking and reactions immediately filtered by my aspiration of being a researcher – in front of myself and possibly others – who did not run any risk of exploiting, silencing, and/or perpetuating the violent hierarchies that allow subordination, exclusion, and exploitation of some – in this case a Syrian migrant man in Denmark?

As an academic I was uncertain of my responses. But I was also uncertain that I was to reply as an academic, in the first place. I could prioritise responses that I would not be ashamed of in front of other academics. I could focus on providing a response that would prioritise Hassan as my companion, but I wasn't sure what that would entail – experiencing anger and compassion the way I would never allow myself to experience in my capacity as a researcher? This late evening interaction left me feeling insufficient and ashamed, both as a researcher and as a person. I felt I had failed, and yet I was unable to understand how.

Failure

There are several ways I can narrate this story in this text through the notion of failure. A growing body of critical theory scholarship argues for thinking about failure as a counter-hegemonic struggle against the neoliberal, racist, heteronormative and universalist structures that set the standards against which one measures oneself (Sjoberg 2018, 87; Halberstam 2011). Within such a framework, failure is being 'leveraged [...] into an analytical lens' and its acknowledgement presented as a path towards more nuanced, care-driven, feminist practices (Laliberte and Bain 2018). Failure, following Halberstam, becomes a political anti-narrative, a form of critique (Halberstam 2011, 88). Thus, the open articulation of failure in critical research is often presented not only as a way of accepting the 'messiness' of the field and the researcher's insufficiencies – thus striving 'for greater reflexivity and honesty in research' – but also 'as a way of refusing to acquiesce to dominant logics of power and discipline' (Halberstam 2011; Harrowell, Davies, and Disney 2018, 231; Laliberte and Bain 2018). In other words, reflexively working through failures is presented as possibly 'productively linked to racial awareness, anticolonial struggle, gender variance, and different formulation of the temporality of success' (Halberstam 2011, 92). And yet, it is precisely the notion of reflexivity as positioned practice, and the articulation of failure as a condition of critique, that my field experience challenged, leaving me in the state of 'impossibility'.

Narrating my experience as failure in terms of academic performance is inadequate; there is something much bigger than scholarly shortcomings at stake in Hassan's story of anger, disappointment, entitlement, and pain. I could recognise that evening as a failure that exposed a series of tensions and contradictions and pushed me towards the unfamiliar and the uncomfortable, towards greater racial awareness, and a more nuanced understanding of the multi-layered workings of systems of oppression and exclusion (Sjoberg 2018, 88; Pillow 2003, 192; Harrowell, Davies, and Disney 2018, 236). I could think through this experience to find ways of problematising the working of racism in a way that would contribute to the possibility of 'coalitional solidarity' across differences (McIntosh and Hobson 2013). I could approach my failure in line with McIntosh and Hobson's suggestion for feminist coalition-building where 'relational failures are inevitable' (2013, 4). And yet I could not. To frame my encounter with Hassan as failure would be disturbing precisely because of the reflexivity the encounter demanded of me, one that I felt Hassan was struggling against. The act of positioning myself such that I am the one who reflects on her behaviour so that Hassan doesn't get hurt makes Hassan yet again the subject of someone else's actions.

Which identity markers matter, and who decides?

It was not so much the racism in Hassan's comments that paralyzed me, as it was his strong rejection of politically correct language. It was not the first or last time during different phases of my fieldwork that I was hearing people of colour with migration experiences fiercely reject political correctness – people whom I knew as otherwise being critical of racism, xenophobia, and nationalist discourses. Here, I suggest reading the repeated discomfort with 'political correctness' not necessarily and/or only as an actual disagreement with anti-racists politics (but also to not exclude such reading). Rather it can be read as an indication of a particular tension that arises from being locked in a position where people's responses are always already a result not just of one's identity markers, but of those identity markers that become highlighted in Eurocentric discourses.

I had trouble responding to Hassan's comments – aware of the various lines of difference between us, I did not wish to come off as patronising and insensitive. But it ultimately meant that I was altering my response to him *because* he was an Arab migrant. And that was in itself not only patronising, but also something he was struggling against – to not be treated differently because of an identity marker that made him particularly visible. Reflecting upon it, I suspect he was rejecting political correctness because he saw it as a practice that made him 'an Arab refugee in Denmark'; a subject of constant racialisation and infantilisation, a subject whose physical appearance provokes and filters reactions. He was, at the end of the day, much more than a Muslim refugee in Copenhagen. He was also a young man in a patriarchal society, a member of the Damascus elite, an Arab and a Muslim in Syria, a country with its own ethnic, religious and social hierarchies.

And yet, all the responses that he faced in Europe, exclusionary as well as those well meaning, would take as their starting point the fact that he is an Arab migrant in a European Union country.

Tension within critical reflexivity

My difficulty in replying to Hassan's comments exposes a particular tension that structures the position of a critical scholar who is committed to, broadly speaking, reflexivity as a means of acknowledging and working through the 'politics of power/knowledge production' with a desire to not reproduce, if not counter, social injustice (Rose 1997).

The tension that I am referring to is, to follow Spivak, the project's 'own impossibility', a moment that indicates the limits but also the possibilities of a given research project. Yet, whereas Visweswaran (1994, 98) argues that 'precisely at those moments when a project is faced with its own impossibility, ethnography can struggle for accountability, a sense of its own positioning', I suggest that those moments of impossibility might also be read as indicative of the impossibility of further encounter (Clifford 1997, 213). Nahal Naficy offers a narration of such impossibility by reflecting on her position as an Iranian scholar conducting research in the US in 2004–2005. She writes: 'in such a turbulent time and place, I found it nearly impossible to conceive of speaking publicly about Iran without being accused of having received either a neo-con Dracula's kiss (if I said anything about human rights abuses or limitations imposed on women) or an Islamic Republic Dracula's kiss (if I said anything about the achievements of women parliamentarians, lawyers, activists, or filmmakers inside Iran, for example)' (Naficy 2009, 115). In the case of my relationship with Hassan, the impossibility arose from a tension between my desire to practice reflexivity as a necessary element of working through the complexities of my fieldwork and the lines of difference that structured my encounters, and Hassan's discomfort with being put in a position where someone engages with him 'reflexively'. In this particular context, reflexivity was an enactment of unequal power relations between me and Hassan. As a researcher, I could not ignore his discomfort, but I also couldn't write about interactions with him without reflexively working through our different positionings.

To frame my inability to respond to Hassan's racism in terms of failure would require me to reflexively engage with the lines of difference between us (legal status, ethnicity, religion) and thus to position each other in a relationship that is based on the recognition of difference. But the difference itself is dictated by Eurocentrism. Recognising the difference means putting Hassan in a position where the response towards him is determined by the same features that make him a subject of racialisation and marginalisation (Finlay 2002, 220). Hassan's rejection of 'political correctness' made me question the practice of reflexivity. This questioning was not targeting reflexivity understood as social critique, as a form of acting the suspicion towards power and knowledge production, or as deconstruction. Rather it posed the question to the very practice of reflexivity as, in its dominant form, a positioned and privileged practice (Pillow 2003, 187–88). A practice that is in most cases tailored for those who are already recognised as privileged within a rather narrow, Eurocentric, and one-dimensional/binary understanding of social hierarchies – white/non-white, citizen/migrant – that leaves little space for 'non-obvious' positionalities to be recognised and guided through the practice of reflexivity.

Human condition

In my hesitation about how to respond to Hassan's racism, I was, despite my best intentions, reproducing him as a 'non-white migrant' before anything else. Moreover, in my desire to act out feminist/critical commitment I was, to quote Patai (1991, 147), running a risk of giving in to my 'own demand for affirmation and validation'. In trying to weigh our oppressions and privileges, I was turning Hassan into a 'project' of my own 'proper' behaviour, a project in which I left much more space for myself to be the wrongdoer than I ever left for him. As I reflect now on my hesitation and on Hassan's anger, I cannot help but read it together with a part of Hannah Gadsby's (2018) stand-up comedy performance 'Nanette', a sharp critique of patriarchy, in which she says, addressing both men and feminists:

> I believe women are just as corruptible by power as men, because you know what, fellas, you don't have a monopoly on the human condition, you arrogant fucks. But the story is as you have told it. Power belongs to you.

Can the rejection of political correctness be read as a convoluted way of requesting recognition of his human condition? As Ravecca and Dauphinée (2018, 133) warn us, we need to be careful to not romanticise social location – 'the "oppressed" are also capable of, and enact, violence' and 'this alerts us to the complexities and contradictions of relationships of domination, which complicate emancipatory politics'.

This is meant as a warning against the tendency to idealise the subaltern and working classes and place in them 'our hopes for political change' (Ravecca and Dauphinée 2018, 133). While analytically, as Ravecca and Dauphinée rightly point out, such idealisation leads to limited and superficial understanding of the working of privileges, oppressions, and violence, in practical encounters, such as the one Hassan and I shared, it can be interpreted as yet another enactment of European racism.

Undoing the European gaze

The impossibility of my encounter with Hassan, and therefore of my project, resides precisely in that I cannot offer an interpretation of this fieldwork encounter, even when narrated as failure, without working through Hassan's and my different positionalities and identity markers. Not working through them would mean silencing crucial power dynamics between our positions in this particular context and thus offering a partial and dangerously naïve interpretation of our encounter. At the same time, working through them would mean placing Hassan, yet again, in a position where he would be recognised primarily through his marginalisation. In that sense, the refusal of political correctness and the insistence on acting out the possibility of being racist can be read as a way of disturbing the discourse in which one is either constantly racialised and framed as a 'dangerous body', or racialised in the reverse manner where his actions are always weighed in a lighter manner because of the marginal position he occupies. Such a reading is the one I am most comfortable with, not only because I find it the most appealing but also because I have limited tools to interpret racism and the rejection of political correctness acted out by those who are marginalised.

This limitation should be read together with a general condition of academic scholarship, including critical theory scholarship, that continues privileging the perspective of the 'ideal' subject (white, European, often male) that defines the limits of knowledge production. I am at a loss when faced with multiple and complex axes of privilege and oppression, as in the case of my encounter with Hassan, where some ways in which I am privileged (white, European) and he is marginalised (person of colour, Muslim, immigrant) are so dominant as sense-making lenses, that other forms of privilege and oppression (Hassan's class background, his gender and the possible entitlement that comes with it, his belonging to the dominant ethnic and religious group in Syria) become hard to grasp.

This is not to say that within the highly racist and xenophobic context of Denmark, one's skin colour and/or religion are not identity markers that make one vulnerable. Rather, continuing to place people within the parameters of the 'white, European, male' gaze offers little space for me to approach my relationship with Hassan, and Hassan's relationship with other 'marginalised' groups, with all their complexities and without a pre-established limit for their interpretation. From such a perspective, Hassan's repeatedly making comments that could be interpreted as racist can be read as his only way of making space for himself in a way that would challenge an otherwise well-established European picture of him as a predominantly racialised person. My reflexive engagement with Hassan's racism was, counterproductively, reproducing me, a white European scholar working with marginalised groups, as the one in the position of power to enact and refuse to enact the violence that Hassan as a non-white migrant was subjected to. Thus, as a concluding remark, by reflexively thinking through my relationship with Hassan and taking the responsibility of being the wrongdoer without leaving much of the same possibility for Hassan, I run a risk of tapping into racist, colonial fantasies of the white man as the one holding access to the complexities of human nature.

References

Alvermann, Donna E. 2002. 'Narrative Approaches'. In *Handbook of Reading Research. Vol III*, edited by Michael L. Kamil, Peter B. Mosenthal, David P. Pearson, and Rebecca Barr, 47–64. London and New York: Routledge.

Clifford, James. 1997. 'Spatial Practices: Fieldwork, Travel, and the Disciplining of Anthropology'. In *Anthropological Locations. Boundaries and Grounds of a Field Science*, edited by Akhil Gupta and James Ferguson, 185–222. Berkeley and Los Angeles: University of California Press.

Finlay, Linda. 2002. 'Negotiating the Swamp: The Opportunity and Challenge of Reflexivity in Research Practice'. *Qualitative Research* 2 (2): 209–30.

Gadsby, Hannah. 2018. *Hannah Gadsby: Nanette*. Comedy performance.

Halberstam, Judith. 2011. *The Queer Art of Failure*. Durham: Duke University Press.

Haraway, Donna. 1988. 'Situated Knowledges: The Science Question in Feminism and the Privilege of Partial Perspective'. *Feminist Studies* 14 (3): 575–599.

Harrowell, Elly, Thom Davies, and Tom Disney. 2018. 'Making Space for Failure in Geographic Research'. *The Professional Geographer* 70 (2): 230–38.

Laliberte, Nicole, and Alison Bain. 2018. 'The Cultural Politics of a Sense of Failure in Feminist Anti-Racist Mentoring'. *Gender, Place & Culture* 25 (8): 1093–1114.

McIntosh, Dawn Marie D., and Kathryn Hobson. 2013. 'Reflexive Engagement: A White (Queer) Women's Performance of Failures and Alliance

Possibilities'. *Liminalities: A Journal of Performance Studies* 9 (4): 1–23.

Naficy, Nahal. 2009. 'The Dracula Ballet, A Tale of Fieldwork in Politics'. In *Fieldwork Is Not What It Used to Be, Learning Anthropology's Method in a Time of Transition*, edited by James D. Faubion and George E. Marcus, 113–128. Ithaca and London: Cornell University Press.

Patai, Daphne. 1991. 'U.S. Academics and Third World Women: Is Ethical Research Possible?'. In *Women's Words: The Feminist Practice of Oral History*, edited by Sherna Berger Gluck and Daphne Patai, 137–154. New York: Routledge.

Pillow, Wanda S. 2003. 'Confession, Catharsis, or Cure? Rethinking the Uses of Reflexivity as Methodological Power in Qualitative Research'. *International Journal of Qualitative Studies in Education* 16 (2): 175–96.

Ravecca, Paulo, and Elizabeth Dauphinée. 2018. 'Narrative and the Possibilities for Scholarship'. *International Political Sociology* 12 (2): 125–38.

Rose, Gillian. 1997. 'Positionality, Reflexivities, and Other Tactics'. *Progress in Human Geography* 21 (3): 305–20.

Sjoberg, Laura. 2018. 'Failure and Critique in Critical Security Studies'. *Security Dialogue* 50 (1): 77–94.

Trinh, Minh-Ha. 1989. *Woman, Native, Other: Writing Postcoloniality and Feminism*. Bloomington: Indiana University Press.

Visweswaran, Kamala. 1994. *Fiction of Feminist Ethnography*. Minneapolis; London: University of Minnesota Press.

10

Researching the Uncertain: Memory and Disappearance in Mexico

DANIELLE HOUSE

My Ph.D. project explored memory and the memorialisation of people currently disappearing in Mexico. I wanted to explore the co-production of memory and place, and how people fight injustice and live among markers and traces of the disappeared. From the outset, then, this project was shaped by impossibilities. How do you research something that isn't understood, even by those living it? How do you research traces, absence, a lack? How do you research memory in a context where the crime is not something of the past but something ongoing? How do you research disappearance when the relatives who search for their missing are murdered? And how do you do so when faced with inadequate methods which seem unable to capture the complexity of the issue, its lived experience, or the insecure context?

Contemporary disappearance in Mexico, at least when I planned and began this project in 2014, felt like an issue obscured by a dark veil. It was a problem that was barely acknowledged, let alone understood, despite the fact that in 2012 the Mexican government released data that showed more than 26,000 people had disappeared (Amnesty International 2013; more recent government data puts the figure at almost 40,000, SEGOB 2018). In reality, this official figure was likely grossly underestimated due to the fear felt by those close to the missing person and bureaucratic barriers that prevent reporting (Open Society Foundations 2016, 4). Many commentators and academics grappling with understanding contemporary disappearances seek to assert an explanation for why it's happening, who is committing the crime, and who the victims are (Gatti 2014, 9). I'm not arguing against trying to understand the causes of the issue. I am suggesting that any analysis that tries to provide or start from simple explanations will be starting from the wrong place. Contemporary disappearance is occurring in the blurred space of the so-called 'war on drugs' (Astorga and Shirk 2010) as well as a long history of state violence (Williams 2011). We can see parts of the structure of impunity and corruption that enable it but cannot explain the issue through a singular strategy or cause. Some disappearances are connected to organised crime, but others are carried out by police officers and the armed forces, and many can be linked to politicians (Al Hussein 2015). The victims are varied: some are targeted, like journalists, environmental defenders, human rights activists, and people with certain skills such as engineers or telecommunications experts (Amnesty International 2013). But many others are seemingly random, and their disappearance had nothing to do with who that person is or what they have done (Calveiro Garrido 2018).

Researching contemporary disappearances in Mexico, then, is framed and shaped by uncertainty. Furthermore, I was looking at social memory and memorialisation which, due to the ambiguous nature of disappearance, couldn't be understood through linear conceptions of time or the prevalent idea that memorialisation can 'deal with' contested pasts (Bevernage 2008; Bevernage and Colaert 2014). It also goes without saying that insecurity shaped and limited what was possible for me to do and know. In this sense, then, this research was always and already a failure; I simply could not fully know or understand the dimensions of the issue at the heart of my project. Yet despite not knowing, not fully understanding, people still act. Relatives of the disappeared are leading criminal investigations into perpetrators and the searches for those missing. Local human rights organisations and lawyers are advocating for legal changes and pressuring the state and the international community. And as we will see, many in society in general are choosing to respond with empathy rather than fear.

This research, therefore, set out to explore (at) the limits of what was knowable, from a position of absolute uncertainty. In this chapter, I describe some approaches through which I carried out the project while accepting these uncertainties and limits. I was never going to fully understand either the issue of disappearance or other people's experiences of it, but the strategies explained below allowed me to gain a complex understanding, to stand alongside, and to glimpse other people's lives and experiences of absence. I want to be clear: I didn't know precisely how I would negotiate these issues before I began my fieldwork, I was unsure what I would spend my time doing. The process was messy, improvisational, and iterative (Cerwonka and Malkki 2007), and I don't want readers to interpret this discussion, specific to my project and context, as a solution to overcoming these problems in general. Moving through the challenges and power dynamics of this research involved a constant consideration of ethics, security, friendship, and politics; of the implications of when and when not to act. I make no claims to success, or to overcoming the difficulties. Instead, in what follows I give three examples of how I negotiated this project, namely: making, listening, and collaborating.

Making

Early on, during my first research trip to Mexico, I was put in contact with two arts-based projects seeking to contribute to public memory and to bring disappearance, and violence in general, into the public consciousness. These projects are Bordando por la Paz y la Memoria (Embroidery for Peace and Memory), an embroidery project which names, on handkerchiefs, the victims of the war on drugs,[1] and Huellas de la Memoria (Footprints of Memory), which engraves and prints the soles of shoes belonging to people searching for disappeared relatives.[2] Bordando por la Paz began in Mexico City in 2011, where a group meets weekly to embroider in public. After going to see them in the plaza one afternoon, I decided to embroider with them almost every Sunday during my months in Mexico. The project had also spread to other cities in Mexico and internationally, and I visited Bordando groups in Puebla and Monterrey. After meeting the artist behind Huellas de la Memoria I got more involved in the project, first translating the shoes into English for a mirror Facebook page, then going to the workshop regularly to engrave and print, and then helping with the project's first exhibition in Mexico City.

It was through participating in these groups, making the collective memorials with my own hands, with these people in these places, that I learnt certain things about how the projects were working, what it meant to create them, and where the value of them lies. Following both Tim Ingold (2012) and Richard Sennett (2008), I see making as a process of thinking. Making allowed me to understand that the process of the becoming of these objects, both materially and socially, is a valuable and rich site of transformation, connection, and knowledge that is in the main overlooked in discussions of memorials which instead focus on the political and cultural life of the 'finished' things. Making – crafting with my hands – made certain experiences and dynamics of disappearance and violence which I was struggling to see, understand, or express, clearer. I also learned a great deal about the issue itself through the projects.

Bordando por la Paz is an effort to sensitise, rehumanise, name, and afford identity to the dead and disappeared. It brings them back into public space and in so doing stitches torn social fabric. One of their goals is to embroider a handkerchief for each person killed in the war on drugs, whether police officer

1 https://www.facebook.com/fuentes.rojas.5.

2 https://www.facebook.com/huellasmemoria/.

or soldier, cartel member, or bystander. Yet they are unlikely to ever achieve this goal; the embroidering and creation, the making, will in all likelihood never catch up with the violence and death, even more so when we think of the insufficient information on the number of victims. Yet I came to realise, through embroidering, that the value of Bordando por la Paz is not in achieving an end goal but is found in the transformation and affect that takes place while making. Through embroidering with Bordando I began to comprehend how slow making with hands is a process that helps people understand their context and their position within it, make social and analytical connections, and find space to construct the community they want and need. There is a sense of achievement once a word has been stitched, a feeling of participation, and a public declaration that these lives are grievable (Butler 2006).

Over the time I spent embroidering I changed; my skill improved, I learnt new techniques, and I made friends. But my relation to the project and how it affected me changed too. When I began to embroider, I felt sadness, desperation, and injustice as I got to know these people through stitching the details of their deaths and disappearances. But as the weeks passed, I spent less time speaking and thinking about the project and the people whose names we were stitching, and more time thinking strategically about what I could do to contribute next. This wasn't a process of desensitisation nor of overcoming and moving on from trauma, but the building of understanding and community.

I first met the artist behind Huellas de la Memoria in a café, and he showed me a pair of shoes belonging to a woman searching for her son which he had just been given. We next met in his workshop, and we talked and engraved and printed a pair of boots. My involvement developed from there. The time I spent engraving and printing the shoes over the ensuing months made me value the material qualities of them, and how these in turn shape the project. Through working with and touching these shoes I saw how they contain stories and speak. I also came to understand, as I watched the project and its collection of shoes grow, that the worn-out shoes of searching relatives reveal some of the spatialities and temporalities of disappearance in Mexico and beyond. The project grew to include shoes that covered Mexican disappearances from 1969 until the present day. They came from across the country, but greater numbers from certain places revealed the epicentres of the crisis, and others were sent to the project from abroad. The growing number of shoes on the workshop shelves materialised disappearance as an issue that was not just contemporary, but which had existed in Mexico with continuity for decades. They showed networks of solidarity and shared experience amongst those living this crisis across Latin America. And the shoes, objects that move, that walk, that march, allowed us to follow their footprints and traces and see the collective tracks that map spaces of disappearance in Mexico.

Like Bordando, the value of this project resides in its detail; in every cut of the lino tool, in every print, in the relationships it cultivates, in the community it constructs, in the pain it conveys, and in the stories it shares and tells. Making and observing in the workshop became inseparable from the conversations we were having and how we shared understandings of memory and disappearance as well as the craft of printing. As with embroidering, talking while doing and making made our conversations richer. I didn't conduct formal interviews with those involved in these projects. This was not because I was unable to – I am certain they would have obliged had I asked. But the only reason I could see to do this was to legitimise my research design. Formal interviews would have added nothing to the understanding that regularly embroidering and engraving with these groups gave me.

I am hesitant to label these activities as any kind of 'method'. This wasn't designed as participatory research, nor was this a classic ethnography. On the one hand, participating in these projects when I was invited to do so was a way to navigate the ethical complexities of this research. But I had also wanted to explore an embodied way of (partially) understanding what it means to be constructing memory in this context, and these projects are deeply about touch, care, community, and connection. Making with and alongside these projects gave me a different knowledge of what they do, they offered a way to act in the midst of uncertainty, and they provided different glimpses of the experience of disappearance.

Listening

A central element in navigating the impossibilities of researching disappearance in Mexico was listening well: listening to what was being said, how, from what context, what was silent, what was repeated often, and what was avoided. This, in many ways, is inherently passive. But attempting to access the everyday experiences of relatives and other activists through interviews wouldn't have worked for several reasons. Ethically (both by my own judgement and that of my university's ethical procedures), I had to restrict my research to relatives of the disappeared who were speaking out about their experiences, were already public figures, and who would not be put at greater risk by speaking to me. Their stories of disappearance, searching, and establishing relatives' associations are generally available online, and I deemed asking for their time to repeat this to me inappropriate. Additionally, I was 'just' a Ph.D. researcher, and I had to be realistic about the limited impact my academic work could have for their cause. Furthermore, I wasn't seeking to find out details of the circumstances of disappearances and searches as such, but what happens after the event of disappearance, how life functions with memory and absence. Formal interviews would not have given me insight into this or the political subjectivity of relatives. Instead, I chose to listen well to what they were saying and doing over time.

I attended seminars, workshops, public events, protests, consultations in the Mexican Senate, caravans, commemorations, and press events. I spent time with relatives, academics, activists, artists, journalists, and others who were dedicated to supporting and working with relatives in their search. I embroidered handkerchiefs and engraved shoes alongside relatives of the disappeared. I listened to what memory might mean as a concept to them, what the sites and locations of memory and absence might be, and how this might be articulated politically. I also wanted to know about more private spaces and practices of memory, as this seemed so important but so absent from the majority of academic accounts of memorialisation. However, with a relatively short amount of time to spend in Mexico, I didn't want to probe deeply into this personal area without first establishing good relationships. Instead, I found books, journalistic writings, testimonials, photographs, and videos to help me learn about how absence is experienced by people in their everyday lives and in intimate spaces. Through listening well, I came to understand the searches that relatives undertake as a practice of memory. Searching for the disappeared is, at its core, a fight against forgetting those who are being erased. Within the search there are sites and material objects of memory – the traces of where that person has been, the personal archives of investigations that are undertaken, sometimes human remains – that fall outside the boundaries of what are considered memory in the academic canon.

In her research on disappearance in Guatemala in the 1990s, Amy Ross (2009, 180) described how, when speaking to relatives, asking certain questions would be enough to convince the person not to respond truthfully as the question itself was foolish in such an insecure context. So, she described, 'Rather than initiating conversations and/or interviews, I listened a lot. I spent years and years with my mouth shut'. Ross undertook what could be confidently labelled ethnography. I listened to roughly one hundred stories of disappearance in person, as well as listening continuously both before and after my time in Mexico through the media mentioned, but I'm not sure I can use that label for my research. Yet Allaine Cerwonka and Liisa Malkki (2007) argue that ethnography is not a methodology in a traditional sense, it cannot be reduced to a standardised technique. Cerwonka writes, 'we stress that ethnography demands a certain sensibility, as well as improvised strategies and ethical judgments made within a shifting landscape in which the ethnographer has limited control' (Cerwonka and Malkki 2007, 20). The complexity of fieldwork never fits the labels of research design, and the labels themselves can be unclear.

For me, this gentle approach to accessing some sense of the experiences of the relatives of the disappeared was the best way, in that context of uncertainty, to navigate my research. It certainly did not produce the 'best work' to demonstrate that I had been rigorous, that I had been a social scientist. Listening gave me glimpses into the experience of living with disappearance and the meanings of memory in that context, while allowing me to negotiate issues of insecurity and ethics.

Collaborating

After some time following and joining in with the projects Bordando por la Paz and Huellas de la Memoria, it became clear to me that I needed to bring these projects and their work to the UK. I care about the issue of disappearance in Mexico and the injustice that surrounds it, and after being welcomed into these projects I wanted to create new spaces for them. Collaboration is complicated and, particularly when connected with the academic career of one person, always exists at a nexus of power, representation, and voice. But there can still be ways to act. Organising exhibitions was one thing I *could* do: I had access to different audiences, I had the English language, and I could create a space to share these projects and stories of disappearance.

Before I made my first research trip to Mexico and connected with Bordando por la Paz, I had already begun to organise Stitched Voices, an exhibition of 'conflict textiles' in the main gallery of the Aberystwyth Arts Centre, with three colleagues from the Department of International Politics at Aberystwyth University (Andrä et al., 2019).[3] Knowing that we were bringing this exhibition together while I was in Mexico and embroidering, it was obvious that I could include Bordando por la Paz in it, should they be interested. There was a year's time lag between the end of my fieldwork and Stitched Voices opening so, rather than rushing commitment to the exhibition, we continued our conversations once I returned to the UK. In the end I borrowed pieces from the three Bordando groups I had developed relationships with: the group in Mexico City I had been regularly embroidering with, the group in Puebla, and the group in Monterrey who were also an association of relatives of the disappeared.

While I was in Mexico City participating in Huellas de la Memoria they held their first exhibition of 85 pairs of shoes. Once I had assisted in and followed this process, we knew that arranging the shoes to come to Europe was a

3 https://stitchedvoices.wordpress.com.

logical next step for the project. This would take it beyond Mexico and I knew logistically what was needed in order to do so. A research group in my department – Performance and Politics International – financially supported the transportation of the shoes, the installation, and the travel for the artist to come to the UK (Edkins 2019, 124). I found exhibition spaces in London and Aberystwyth, and the Huellas de la Memoria Collective fundraised to enable the mother of one of the disappeared Ayotzinapa students to join. I was supported by countless people and organisations in London and Aberystwyth to realise the exhibition and a range of talks and activities. The Collective in Mexico used their networks across Europe to find people who would organise an exhibition of the shoes and prints in their cities so it could tour. In the end it moved across France, Germany, Italy, the Netherlands, Belgium, and other countries for several months.

These exhibitions were not the sort of public engagement activity that we, as academics, are pushed to deliver; they were not promoting my work or research 'findings'. The intention was to share these projects and these issues with an audience that may not have realised disappearance was taking place in Mexico, and for a Mexican audience to see that people elsewhere in the world were paying attention. That the exhibitions were possible at all was due to the relationships, trust, and intimacy we had built, but I still worried about the ways their projects and objects, and the issue of contemporary disappearance in Mexico, were represented in the exhibition narratives, spaces, and accompanying events. I was also concerned with how the shoes, prints, and handkerchiefs were treated, for example, when in transit and during installation. From my time working on these projects I knew the emotional connection between the object and the maker; the value they had went far beyond monetary. Through regular communication on decisions and plans, we prioritised what was and what was not important to those who had lent me things, at times counter to what I had assumed. For good and for bad, we negotiated what it meant to bring these objects, created in protest, solidarity, and intimacy, into the different exhibition spaces.

When writing about her experiences of working with the Sangtin Writers in Uttar Pradesh, India, Richa Nagar (2006, XXXIX) explained that when, as the academic in the group, she became anxious about the power of representing the other women and their journey, the group reminded her they had formed an alliance, 'strategically combining, not replicating, our complementary skills'. Acknowledging the inescapable ethics and power dynamics of academic and activist collaborations means that any attempt to negotiate collaborative work is risky. For these exhibitions, it wasn't necessarily the uncertainty of the issues and research context that created the conditions for failure, but the uncertainty of the collaborations themselves: they didn't have guarantees, they could have fallen apart, many things could have gone wrong. There was, therefore, a risk of 'doing something'. But I wasn't at any point doing these things alone. The risk was shared, and we acted together in uncertainty. I'm not sure the process of organising these exhibitions revealed something to me about disappearance, but it did force me to face the complications of 'doing' memory; questions of representation, curation, voice, and the politics of space.

Finally, although I am proud of these exhibitions, prioritising them alongside the doctoral thesis came with downsides. In the context of the neoliberal university, using my time to produce these came at the expense of gaining teaching experience, attending conferences, and working on publications. I am happy with the choices I made, but as early career researchers we face the unrealistic expectation that we can do everything, that we can undertake fieldwork based in close relationships that take time, energy, and resources, while simultaneously delivering publications, public engagement, and teaching. Since completing the Ph.D., applying for jobs has presented another issue to navigate: commodifying these exhibitions, and so inherently my relationships and other people's experiences, on my CV, to demonstrate my experience of collaboration, public engagement activities, and creative methods.

Conclusion

My project, from the outset, was framed by failure: it was impossible to understand, both conceptually and literally, its central issue. As I explained in the introduction, working on contemporary disappearance in Mexico has felt at times like scrambling in the dark. Alongside fear, the goal of disappearance, as a crime and a political act, is an ongoing and ever-present absence – of information and of persons. The additional issues of insecurity, a lack of data, and a focus on memory, pushed my research to the limits of the knowable. Instead, I tried to accept and work within uncertainty. I have not set out to offer my experience as a guide for how to 'do' this kind of research, but instead wanted to acknowledge and explain the challenges of it and share how I attempted to negotiate them. Some of the issues I have briefly discussed are universal to fieldwork and research – the ethics of researching others and attempting collaborations for example – but other issues were specific to the problem of ongoing disappearance in this context. This chapter discussed the particular insights and knowledge gained from approaching research in this way. Making, listening, and collaborating enabled specific glimpses into the everyday political, social, and emotional impacts of disappearance and memorialising the disappeared.

* *The author would like to thank the editors for their constructive and insightful*

comments and guidance on this chapter and also thank and acknowledge the work of the various collectives whose work and projects are drawn on in this chapter: Fuentes Rojas, Bordando por la Paz Puebla, FUNDENL, and Huellas de la Memoria. I lastly want to thank London-Mexico Solidarity, Performance and Politics International, and the many other people, colleagues, and organisations who supported and helped realise the exhibitions and associated events.

References

Al Hussein, Zeid Ra'ad. 2015. Statement of the UN High Commissioner for Human Rights, Zeid Ra'ad Al Hussein on his visit to Mexico. 7 October 2015. *Office of the High Commissioner for Human Rights*. http://www.ohchr.org/EN/NewsEvents/Pages/DisplayNews.aspx?NewsID=16578.

Amnesty International. 2013. *Confronting a Nightmare: Disappearances in Mexico*. London: Amnesty International. http://www.amnesty.org/en/library/info/AMR41/025/2013.

Andrä, Christine, Berit Bliesemann de Guevara, Lydia Cole & Danielle House. 2019. 'Knowing Through Needlework: curating the difficult knowledge of conflict textiles'. *Critical Military Studies,* Online First.

Astorga, Luis, and David A. Shirk. 2010. 'Drug Trafficking Organisations and Counter-Drug Strategies in the U.S.-Mexican Context'. *USMEX WP* 10–01: 1–47.

Bevernage, Berber. 2008. 'Time, Presence, and historical injustice'. *History and Theory* 47 (2): 149–167.

Bevernage, Berber, and Lore Colaert. 2014. 'History from the Grave? Politics of Time in Spanish Mass Grave Exhumations'. *Memory Studies* 7 (4): 440–456.

Butler, Judith. 2006. *Precarious Life: The Powers of Mourning and Violence*. London and New York: Verso.

Calveiro Garrido, Pilar. 2018. 'Desapareciones: de la llamada Guerra Sucia a Ayotzinapa'. In *Cartografías Críticas Volumen I: Prácticas políticas que piensan la pérdida y la desapareción forzada*, edited by Ileana Diéguez (complication of texts), Paola Marín and Gastón Alzate. Los Angeles: Ediciones KARPA. http://www.calstatela.edu/al/karpa/cartograf%C3%ADas-cr%C3%ADticas-volumen-i.

Cerwonka, Allaine, and Liisa H. Malkki. 2007. *Improvising Theory: Process and Temporality in Ethnographic Fieldwork*. Chicago and London: The University of Chicago Press.

Edkins, Jenny. 2019. *Change and the politics of certainty*. Manchester: Manchester University Press.

Gatti, Gabriel. 2014. *Surviving Forced Disappearance in Argentina and Uruguay: Identity and Meaning*. New York and Basingstoke: Palgrave Macmillan.

Ingold, Tim. 2012. *Making: Anthropology, Archaeology, Art, and Architecture*. London and New York: Routledge.

Nagar, Richa. 2006. *Playing with Fire: Feminist Thought and Activism through Seven Lives in India*. Minneapolis and London: University of Minnesota Press.

Open Society Foundations. 2016. *Undeniable Atrocities: Confronting Crimes Against Humanity in Mexico*. Executive Summary. New York, NY: Open Society Foundations. https://www.opensocietyfoundations.org/sites/default/files/undeniable-atrocities-execsum-eng-20160602.pdf.

Ross, Amy. 2009. 'Impact on Research of Security-Seeking Behaviour'. In *Surviving Field Research: Working in Violent and Difficult Situations*, edited by Chandra Lekha Sriram, John C. King, Julie A. Mertus, Olga Martin-Ortega, and Johanna Herman, 177–188. Abingdon: Routledge.

SEGOB, Secretariado Ejecutivo del Sistema Nacional de Seguridad Pública. 2018. *Registro Nacional de Datos de Personas Extraviadas o Desaparecidas (RNPED)*. https://www.gob.mx/sesnsp/acciones-y-programas/registro-nacional-de-datos-de-personas-extraviadas-o-desaparecidas-rnped.

Sennett, Richard. 2008. *The Craftsman*. London: Penguin Books.

Williams, Gareth. 2011. *The Mexican Exception: Sovereignty, Police and Democracy*. New York: Palgrave Macmillan.

Part IV

Writing as Translation

11

What Might Have Been Lost: Fieldwork and the Challenges of Translation

RENATA SUMMA

'I will soon go back home and I am thin, exhausted, a lint; lean but tanned. I am thin due to coffee, beer and rakija, to the smoky rooms and that language, that pops in my ears, that is lost in my mouth and which tells me that it is possible before it denies me any sort of understanding. I am thin due to the sleepless nights. How many? All of them. Due to the prayers I hear when the day rises. I live out of air, caffeine and curiosity. I live to always ask for more, and always receive more. I live in the slow pace of the crowded trams, tirelessly walking back-and-forth through the narrow and steep streets, trying to cover every inch of this city, trying to grasp it all. I live out of sunshine, yes, when there is some. Out of wind, yes, when there is some. I live out of stories and wounds, so many of them. I live out of songs, bells and dogs barking through the night; out of the smell of clean sheets, sweaty scarfs and sugary smoke. Burnt firewood, forest, river and jasmine. I live out of these eyes, thousands of them, and their improbable colours. And by living like this I lose myself. I lose weight, nights of sleep, appetite, purpose. This city seems to escape me.' (3 May 2015)

This extract from my field notes, written a month before I had to leave Sarajevo on my second research trip for my Ph.D. thesis, illustrates two main concerns that I had during my research, and which will be at the centre of this chapter. The first one, which I discuss in the following section, is the attempt to 'capture the city'. Indeed, the text above exposes the anxieties of realising the difficulty of understanding 'what is really going on' in Sarajevo and the desire to blend in the everyday of the city to 'grasp it'. Although I had already reflected on the problems of treating the fieldwork as raw data that the researcher can collect to arrive at a 'true account' of the situation on the ground, I still could not completely overcome those scholarly assumptions that isolate 'theory' from the 'real world' and distinguish the subject from the object.

The second concern is the multiplicity of experiences which comprise the fieldwork. How do we translate the myriad of narratives, interviews, opinions and the variety of life stories into coherent social analysis? How do we translate the sounds, rhythms, emotions, coffee breaks, friendships, breakdowns, and failures into a cohesive text? How do we translate the inconsistences of the everyday life of fieldwork into knowledge? Those are the questions that I will tackle in the second part of this chapter.

The quest for legitimacy: Trying to capture Sarajevo

Sarajevo has frequently made me feel like an outsider, if not a fraud. I moved to the capital of Bosnia and Herzegovina in the middle of my Ph.D. to conduct research in everyday places both in Sarajevo and in Mostar, to understand how ordinary people enacted and displaced ethnonational boundaries that were institutionalised by the Dayton Peace Agreement. The encounters I had with the 'local population' and with other researchers frequently made me uneasy and questioned my legitimacy as a researcher. Indeed, I have never fallen into the most frequent categories one thinks are necessary for a researcher to get interested in Bosnia and Herzegovina (BiH). I am not from BiH nor from the Bosnian diaspora. I do not come from the 'region', i.e., ex-Yugoslavia, nor from Western Europe, which has been naturalised as an actor who should look at the fate of BiH as their responsibility (Brljavac 2011). Instead, I come from Brazil, a country which made my interlocutors raise their eyebrows every time I mentioned it.

Why would someone from Brazil be interested in Bosnia? I have tried to answer this question frequently, searching for alternative and better explanations each time someone asked it. Usually this question was raised while I was in the company of other researchers, mostly Europeans, and along the following lines: 'I understand why an Italian or a German would be interested, but… Brazil? This is so random'. Even though those questions reveal and reproduce a geographical epistemic imaginary that I refute – who can produce knowledge and about what – it is needless to say that such comments only reinforced my feeling of non-belonging. They were a constant reminder that I was an outsider in BiH.

Indeed, the production of knowledge in academia relies not only on what is being stated but also on who makes the statement. Legitimacy to make statements is, thus, a fundamental pursuit of the researcher's life and conducting fieldwork is perceived by many as a way to achieve it. After all, it is frequently argued that fieldwork is the moment when the researcher goes out of their ivory tower to meet the 'real world', to make a truer encounter with the object of their research. This presumed direct contact with the object – in opposition to the mediated contact provided by, for example, written work on the matter – would give the researcher, one assumes, an additional, more exclusive and complex knowledge of the object. Therefore, fieldwork is frequently mobilised as a legitimising experience: one has been there, thus one knows and is entitled to talk/write about it.

This chapter argues that knowledge produced during the fieldwork is already mediated and interpreted. Moreover, it suggests that there is a preceding step one has to face even before being able to claim this (often contested) status, and that questions the very (il)legitimacy to conduct fieldwork in the first place. I argue that, in my experience, my outsider status was the reason behind the quest for 'capturing the city'. If the subject cannot make proper sense of the situation or is not authorised to produce knowledge about it, then the object would have to play a leading role in revealing its own true nature. Nonetheless, such a statement stands in sharp contrast to what I understand about how research and knowledge are produced, i.e., that the researched object 'holds no ontological status apart from the many and varied practices that constitute their reality' (Campbell 1992, 11); or, according to Foucault (1984, 127), 'we must not imagine that the world turns toward us a legible face which we would only have to decipher; the world is not the accomplice of our knowledge'. This is why it is so difficult to 'grasp the city'.

Moreover, my positionality did affect my research more than I had anticipated. I had already foreseen that being a young female researcher would probably constrain some opportunities during this period. I had heard about how male researchers are usually taken more seriously than women. Still, it would have been useful to engage more deeply in discussions about how gendered power relations could disrupt relationships of trust that had taken quite some time and effort to build, limiting the possibilities and paths the research could take. However, I had not anticipated that other features would be so important while conducting my fieldwork, and I mean not only the fact of being a foreigner but also – and more specifically – the place I come from.

Let me be clear here. I acknowledge that the place I come from was less a 'real' obstacle to getting interviewees, and more a source of doubting the legitimacy of my presence there. Sometimes, it actually helped me getting access since most people thought I was 'exotic' and seemed more open to discuss 'delicate issues' with someone who was understood as more neutral.

Although I had prepared myself to conduct this work – by reading extensively about my research topic, taking 'Serbian/Croatian' language classes and having actually been there a few years prior to that – the overwhelming feeling that overcame me as soon as I landed in Sarajevo was one of being a stranger, of having so much more to learn than to state, of looking around, and doubting if I were making *proper* sense of the situation.

I conducted fieldwork in Sarajevo three times from 2014 to 2015, in a total of almost six months, to try to make sense of how ordinary people enacted ethnonational and international/local boundaries in their everyday lives in a post-conflict society. One of my first steps when I got there was to take a bus to the neighbourhood crossed by the Inter Entity Boundary Line (IEBL), an official boundary drawn by the international community during the Dayton Agreement's negotiation to divide Bosnia and Herzegovina into two administrative entities (the Federation of Bosnia and Herzegovina, shared among Bosniaks and Croats, and the Republika Srpska, led by Serbs) – a site where I would spend a significant amount of time doing research. It was a December afternoon, a usual Friday. There was snow on the ground and virtually nobody walking on the sidewalks in this suburban area, just a few cars rushing from one side to the other. I had walked across/through dozens of empty blocks, closely paying attention to all the details such as street names, official signs, dwellers' names on the block entrances, graffitis on the wall among others, trying to understand what it meant to live there, how it felt to live in a former war front which was transformed into two distinct neighbourhoods. After walking back and forth a few kilometres, I went back to my place not only with frozen hands and nose, but also with the feeling of impossibility. After all, there was *nothing really going on* there or, if there was, I could not understand what it was. The city seemed coded, locked and secretive. Walking was one of the practices I adopted to try to 'open up' the city in order it to reveal its secrets to me.

That is why I was, in my 'spare time', 'tirelessly walking back-and-forth through the narrow and steep streets, *trying to cover every inch of this city, trying to grasp it all*'. De Certeau (1984) warned us about the meanings of walking the city, which surpass the division between the researcher and the object of the research. Indeed, when one walks, one 'writes' the city as well and becomes a practitioner by making

> use of spaces that cannot be seen [...] The paths that correspond in these intertwining, unrecognised poems in which each body is an element signed by many others, elude legibility [...] The networks of these moving, intersecting writings compose a manifold story that has neither author nor

> spectator, shaped out of fragments of trajectories and alterations of spaces: in relation to representations, it remains daily and indefinitely other (De Certeau 1984, 93).

Thus, the city we walk is already a city we are producing, by occupying its spaces, connecting its arteries, attributing meanings to its places. It is a city where the 'walker actualises its possibilities, by making it exist' (Ibid., 98), while also inventing other patterns and creating new meanings.

My field notes, however, reveal that my concerns of that time went in different directions. I still kept trying to capture the *truest* meaning of boundaries enacted in the city. I was still wondering if there was a more *real* Sarajevo than the one that I was experiencing. Behind those concerns, lay implicitly the idea that the city had an essence which I, as a researcher, was responsible to grasp, to expose, to translate into academic knowledge.

> I wake up sweaty and decide to extend my stay for another three nights, almost four days. One way to avoid thinking that, now, I have only a month left. In fact, it is only three weeks [...] (Field notes, 7 May 2015)

All the elements presented above – the feeling of non-belonging (and, on some days, of being a fraud), the uncertainty of attributing meanings to people's actions, the pressure to understand what was really going on in the city – gave me the impression that the results of my work were never enough. I needed to stay longer, to walk more, to carve out new information. The following questions haunted my stay in Sarajevo: How long is long enough to actually capture the dynamic of a city? How many days do I need to be there to be taken seriously, how many interviews do I have to conduct to produce relevant knowledge, how many kilometres do I need to walk before the strange becomes natural?

By the end of my last research *séjour*, I came to accept that my attempts to 'grasp' the city were impossible. However much I could try to cover, I would always fail to grasp it. No matter how many voices I heard to try to overcome my limited subjectivity, my work would always be subjective. No matter how closely I looked at Sarajevo, the city would always escape. Although I had attended many research seminars and have read about post-empiricist and post-hermeneutic modes of inquiry (Shapiro 2013, 21), it was only by failing to capture the city that I have finally understood the impossibility of 'collecting data' in order to grasp any essence of Sarajevo beyond the many everyday practices, experiences, encounters, and stories that I had lived or heard during my fieldwork.

The difficulty of translating fieldwork

When I went back to London, where I was doing part of my Ph.D., the first question my co-supervisor asked was 'what contribution did the fieldwork make to my thesis'. I froze and could not formulate an answer right away. I thought about this rich period when, after almost two years locked in the library, my work was, for months, mostly to talk to people and to pay close attention to the city's everyday – again. I had interviewed dozens of people, made a few good friends, experienced an earthquake for the first time, been chased by stray dogs, been exposed to both the saddest stories and the famous Bosnian sense of humour; in short, I had lived tirelessly all those months. My memories of those periods are full of emotions, sounds, scents, and contradictions, all of them hardly translatable into academic knowledge. How could I connect them to the established body of literature I was relaying on? How could I do justice to the plurality of life to which I was exposed during this period?

Since my main interest in conducting fieldwork was to understand how ordinary people enacted everyday boundaries in contemporary BiH, one of greatest challenges was to translate the multiplicity of approaches my interlocutors had regarding their enactments and displacement of boundaries in those two post-conflict cities into a cohesive account, or at least an account which did not contradict itself. Dealing with such a large amount and variety of information gathered often around the same topic – often a disputed one – and transforming it into knowledge is perhaps one the biggest challenges of conducting this kind of work.

Dauphinée (2013) has brilliantly explored an important aspect of this challenge. In her book 'The Politics of Exile', while she describes the semi-fictional encounter between an American professor who works in Canada and 'who has built her career on the war on Bosnia' and a veteran Serb who had moved to Canada after the war, she illustrates the different meanings attributed to a specific event according to one's position. Stojan, the war veteran's character, disagrees on the periodisation of the Bosnian war as it is established in history books. While the official narrative states that the war started on the 6 April 1992, Stojan disagrees: 'For me, it was something very different than what you described [...] it was unclear when it began, [...] for example, I was just at home in Pale at the time you say it officially started. And I didn't know it had started' (Ibid., 54). War only came to him the day his brother got recruited to join the newly instituted army of Republika Srpska (VRS). Republika Srpska is a territory within Bosnia and Herzegovina that unilaterally declared independence in January 1992, constituting, at the same time, its own Army (VRS). This event is usually considered as one of the main steps leading to war in BiH (1992–1995).

In Dauphinée's book, the encounter with Stojan's character evolves into a situation in which the professor questions everything she has ever written on Bosnia, and as a result destroys all her work. This might be a strong image which Dauphinée employs. However, information one comes across during fieldwork and which do not corroborate or do not fit into one's own hypothesis or the more established narrative might provoke a similar research earthquake. Coming into contact with dozens or even hundreds of different points of view might turn the task of generalising or trying to produce a coherent narrative of the researched subject very difficult. Thus, although Dauphinée's book is more about the ethics of research and the (im)possibility of love as ethics, it also helps readers think about how different positions and experiences – in that case, during the war – get translated (or not) into academic knowledge.

This was an important issue for my research due to the multiplicity of positions that I had encountered during fieldwork. Indeed, the interviews I conducted focused on my interlocutors' own experiences and impressions of their everyday lives in post-Dayton Bosnia and Herzegovina. Not rarely, the responses I received were quite contradictory vis-à-vis each other. This is not so surprising: I was not asking them about the official version of the facts – even though those are also deeply disputed among different communities inside and outside BiH – but about their own experiences. However, that posed a challenge, again, of how to translate a multiplicity of positions into an understandable construction of academic knowledge. One way of making this more transparent is to adopt a written strategy and style which is not totalising, i.e., which exposes the fragmentary character of the production of knowledge. Acknowledging the messiness and complexities of the situation allows for not dismissing the plurality of narratives while fixing a single story.

One way to tackle the challenges of translating fieldwork is through the writing style. In my thesis, I thus adopted 'snippets', 'small fragments inserted on the narrative in order either to illustrate some points or to allow for further analysis, as a method of presenting the research' (Summa, 2019). Because everyday life is much messier than what is stated by official documents, this heterogeneity of practices and voices needed to be made explicit. The 'snippets', therefore, stressed the precariousness of the order in which everyday life is embedded, revealing, instead, the cracks through which alternative spatiotemporal practices may emerge. Therefore, using 'snippets' affects the writing and the ways of presenting the text: it sometimes moves from one place to the other, bringing different voices on the subject and creating a heterogeneous picture of the places I had explored.

However, this method does not solve the problem entirely and could never fully reproduce the movement of the field into a written text. In writing, we inevitably freeze our interlocutors' experiences and turn what it is dynamic into some degree of stasis. That became clear when, after completing my Ph.D., I went back to Bosnia and Herzegovina to give some feedback to my interlocutors there. Since I had interviewed many people during my fieldwork, I decided to look only for those whom I considered to have given bigger contributions to my thesis. One of the first challenges was finding them again. I went back to some of the everyday places that figured in my thesis only to find that many things had changed: the barman I had interviewed had been laid off, the NATO forces who were very keen to talk to me and showed me around their military base had been deployed to other missions or returned to their home countries, some friends had left the country along with thousands of young Bosnians from all ethnonational belongings to find better jobs in Western Europe. While I had been locked in the library for a year finishing the thesis, the lives of those who populate my work had shifted, and so the everyday places that were central to my analysis had changed. I kept wondering how different the stories would be – how much had the city changed? Going back to Bosnia and Herzegovina was the final reminder that the everyday always escapes: the city is in permanent movement and it is only possible to be grasped punctually and for a very short period of time.

Conclusion

This chapter does not exhaust all the difficulties of conducting fieldwork. Through a reflection on my fieldwork experiences in Bosnia and Herzegovina, I suggested two main concerns that preoccupied me while I was conducting my fieldwork in BiH, and while I was trying to make sense of it back in Rio. The first of them regards the insistence and, at the same time, the impossibility of discovering a more real city that, for a long time, seemed coded to my eyes. My days were haunted by the impression that the city always escaped my attempts to grasp it and the sensation that so much had been lost: in translation, in the stories I had not heard, in the people I could not interview, in the spaces I could not get access to – but also in lazy weekends when I would prefer to go out with my friends instead of going to the outskirts of Sarajevo once again. The second concern came to the fore when I had already left BiH and had to deal with this heterogeneous corpus of 'things' – interviews, impressions, thoughts, pictures, emotions, memories – that composed but not exhausted my fieldwork experiences. How could I turn them into academic knowledge? Again, my impression is that so much got lost along the way.

Neither of those points are about turning failure into a productive or desirable experience. However, having struggled with both of those questions, I came to wonder if the fieldwork is not, in fact, about the possibilities and impossibilities of translation. If there is not a 'more real city to be grasped', the only stories that are possible to be told are the translations of ours or our interlocutors' experiences located in particular places and in particular times.

References

Brljavac, Bedrudin. 2011. 'Europeanisation Process of Bosnia and Herzegovina: Responsibility of the European Union?'. *Balkanologie* 13 (1-2).

Campbell, David. 1992. *Writing Security. United States Foreign Policy and the Politics of Identity*. Minneapolis: University of Minnesota Press.

Dauphinée, Elizabeth. 2013. *The Politics of Exile*. New York: Routledge.

De Certeau, Michel. 1984. *The Practice of Everyday Life*. Berkley: University of California Press.

Foucault, Michel. 1984. 'The order of discourse'. In *Language and Politics*, edited by Michael Shapiro, 108–38. Oxford: Oxford University Press.

Shapiro, Michael. *Studies in Trans Disciplinary Methods. After the aesthetic turn*. New York: Routledge, 2013.

Summa, Renata. 2019. 'Inventing Places: disrupting the divided city'. *Space and Polity* 23 (2): 140–153.

12

I Don't Know What to Do with Myself: 'I' as a Tool, a Voice, and an Object in Writing

KATARINA KUŠIĆ

Fieldwork sets us up for a myriad of writing failures: not 'doing justice' to our topic, not representing our interlocutors 'faithfully', obscuring ourselves in a method that is inevitably embodied, immobilising in text something that is always fluid. This chapter dwells in the many anxieties that underpin writing failures by interrogating the use of 'I' in writing up our fieldwork in International Relations. In an attempt to *show* the different uses of myself in writing, the chapter consists of three sections. The first section is cut from the first full draft of my Ph.D. thesis.[1] The second part is the same section rewritten for the version of the thesis submitted for examination. I leave the two unedited as illustrations – even if I would now write them quite differently. The third part of the chapter discusses the very different uses of the 'I' in the pieces. Specifically, I interrogate my reasons for deciding *against* a more narrative style of writing – a decision that contradicts my epistemological and political commitments.

The cuts: From the first draft of the Ph.D. thesis

Fieldwork journal, Centre for Cultural Decontamination (CZKD), *School of Knowledge: Srebrenica, mapping genocide, and post-genocide society*.[2]

I could write an ethnographic story about today's event on education about

[1] Submitted to the Department of International Politics at Aberystwyth University.

[2] CZKD is a cultural non-profit institution famous for its opposition to the regime of Slobodan Milošević and its nationalist policies in the 1990s, and working on highly politicised issues of transitional justice. Belgrade, 3 March 2016.

Srebrenica. I could write about the fact that only 20 people clicked 'attend' on Facebook, meaning those who showed up are still reluctant to make themselves so public and vulnerable to harassment. I could write about the comments on the event being about the usual conspiracy theories about Soros's[3] money and the wasting of resources on talking about the past. I could talk about the fact that, even though I've been talking about how nationalism is totally passé these days, the event still had to have police presence in front of CZKD. I could write about my friend warning me to expect inappropriate comments in the end, because that 'always happens at events like this'. I could write about the uncomfortable feeling of not only disgust and anger – fuelled by a film detailing the timeline of the Srebrenica genocide, but also the boredom caused by the impression that we are always talking about the same things, and wondering if this film will really benefit any one in any way.

This would make an excellent ethnography about the Balkans – complete with genocide and the simmering hatreds. But none of this really made an impression on me – I sat there, mildly interested, watching people around me and wondering whether different generations present have different reasons for attending – some of these people must have attended the anti-war protests in the early 1990s in Belgrade, while others were born into the 'post-genocide society' we are discussing this evening. None of this really impressed me, until Dr. Svjetlana Nedimović from Sarajevo started speaking. She looked distant and bored while others were speaking, and she started in a style that I deemed a bit too emotional: saying that it's hard for her to come to Belgrade for the first time, and explaining that she has a difficult relationship with a city she has never visited before. She also explained that she is reluctant to use the label of a political activist, but that she does 'street politics' [*ulična politika*] these days.

She started the actual paper with an anecdote that quickly transported us from genocide timelines to contemporary Bosnia and Herzegovina (BiH). She was protesting the agreement to build a dam in a national park in BiH with a group of activists – 'how can someone who has a "miserable four-year mandate" [*mizeran četverogodšnji mandat*] think that he has the right to decide on a destruction of a national park?' They were standing in front of the office where the agreement was signed, the signature marking their defeat. In the discussion on what to do as the politicians were exiting the office, some of the activists had the idea to do an ironic applause. Svjetlana did not approve – irony has no place in this matter, except admitting defeat. As the politicians

[3] Here, I refer to the anti-Semitic conspiracy theories about the philanthropic donations of George Soros. These theories have been popular in Serbia and the region more generally for more than a decade but are now quite common in the US and the UK as well.

were walking out, she took out a rotten tomato from her bag and threw it at one of the people. She missed, but heard screams nevertheless – it was not her or the guy she aimed at screaming, it was her fellow activists. The screams were because she resorted to violence, violence that is unjustified, impolite, and unproductive.

Her first instinct was to attribute this to the 'post-genocidal' state of her society – 'we are so traumatised, that we consider a tomato a violent act. But this would be too easy, this would not take into account our contemporary experience. And our contemporary experience is one in which the political act is defined by the EU and other international agencies in Bosnia and which cannot stand direct action'. She went on to deliver a paper about how the BiH people are the way they are (with constrained ideas of political life that cannot handle a tomato) not only because of war, but also because of the social and political order that they built in the past 20 years under the direction of international agencies.

[...]

Ethnography as text is necessarily representation, rather than reality captured: sitting in CZKD, I was aware of my power to write culture. I knew that I could represent Serbian society through these observations, and I also decided not to. My own position here is ambiguous: as a consumer of academic texts, I know what boxes to tick to make my representation 'a good ethnography of the Balkans' – some pain, suffering, and war are easily employed to 'draw in' the reader. However, I am also from the region – growing up with these issues somewhat removes the spectacle, I feel *bored* and in turn I feel *guilty* for that.

The submission: Excerpt from the submitted thesis

This not only reduces ethnography to a 'data-collection machine capable of accessing unmediated reality in all its authenticity and accuracy' (Vrasti 2008, 281), but also resonates with other sources of experience. Namely, it is easily mixed with the idea that we similarly *know* because we are somehow *from* a specific region – a fact which is also supposed to give us some unique, privileged access. Without denying the importance of context, I want to argue here that the main contribution of an ethnographic approach is *not* its privileged access, but its possibility to reflect upon issues of that access – the idea that no matter if we are at the desk, or immersed in the field, we always *know from a specific location*. This is what Jon Harald Sande Lie defines as anthropological reflexivity – 'the constant and reciprocal relationship between fieldworker and informants, underscoring that the fieldworker's position in the field influences the data that she gains access to and acquires' (Geertz 1973; cited in Lie 2013, 205).

Here, the above mentioned 'crisis of representation' becomes more than doing away with 'the field' as an adventurous voyage to the 'other' lands – it also instructs us to pay attention to ways in which we are not only biographically, but also historically and socially connected to the sites we study (Gordon 1997; cited in Gupta and Ferguson 1997b, 38). Instead of making ourselves disappear, a task impossible even with the best intentions, we are invited to reflexively consider these connections, how they influence the researcher, the researched, and the research itself. In my case, these connections are even more prominent as I grew up in what is today Croatia, in a region that shares a border and war history with Serbia. I travel with a Croatian passport that has free access to the EU (unlike a Serbian one), and my Croatian accent and dialect display my origin in every conversation I have in Serbia. However, while these facts made me a foreigner in Serbia, my language familiarity, the ability to read Cyrillic, and cultural proximity, marked me as one 'studying their home' among colleagues in the UK – in the eyes of British scholars, I was doing 'native ethnography'.[4]

[...]

The last site for situatedness is the text itself.[5] Cecilie Basberg Neumann and Iver B. Neumann (2015) offer two ways of going about this: one is the reflexive wager which produces a text that is centred on how the researcher herself is changed during the research process. Dauphinée's *Politics of Exile* (2015) is probably the most famous example. The second option is the analyticist's way – in this option, we do not focus on ourselves to interrogate the structures that make us and the world, we do not rely on introspection. Instead, an analyticist 'focusses on coming to terms with themselves as an instrument of data production' (Basberg Neumann and Neumann 2015, 815). The difference is also described as one between a *methodological* situatedness that is dealt with analytically, and a *methodic* situatedness that is dealt with reflexively: 'Where a reflexivist researcher tends to handle the relation between the interlocutor and researcher by asking how interlocutors affect them, an analyticist researcher tends to ask how the researcher affects the interlocutors' (Basberg Neumann and Neumann 2015, 817).

---

4 The idea of 'native anthropology' was one of the products of the 'crisis of representation'. By 'repatriating' anthropology and studying and denaturalising 'home', instead of the Other, anthropology was meant to avoid practices that exoticise its subjects.

5 For a great overview of how anthropologists have handled 'writing themselves in', see the contributions in *Tales of the Field* (Van Maanen 2011).

Following this distinction, this thesis falls into the analytical camp – I think of 'method as a question of producing data by bringing certain value commitment with [us] into the field' (Basberg Neumann and Neumann 2015, 818–19). Though my sense of self influences the research, and the research process influences my sense of self, I do not make space for this in a thesis that is ultimately not concerned with *me*. However, I also do not aim to erase myself from the text: it was me who read theories, spoke to people, thought about questions, and analysed observations. I make this visible in the text not to limit its scope, but to practice the ethnographic stance that emphasises that all knowledge is produced from a specific *location* – to practice 'strong objectivity' by putting subjects and objects of research on the same 'critical, causal plane' (Harding 1992, 458). In doing so, I do not shy away from making knowledge claims, but make them with a claim to *strong objectivity* (Harding 1992).[6]

Discussion

The first draft of the thesis saw each chapter open with a short vignette like the one in the first section above. I wrote them because I wanted to make myself *visible* even in the chapters that talk about more technical and conceptual aspects of my thesis: statebuilding, intervention, the local turn, and governmentality. I wanted to show the importance of lived experience for my own thinking: the thesis itself argues that lived experience is crucial for rethinking IR concepts. I decided to cut them out after a conversation with my supervisor, we made a joint decision to 'play it safe' and 'keep the ethnography to the ethnographic chapters'. Despite cruelly discarding them, I liked my vignettes – I stored them in one of the many 'scrap' files I kept, this one titled 'stories'.

When comparing the vignettes with the finished (submitted) product, I was jolted by the difference. Instead of relying on lived experience to explain the conceptual choices and biographical connections that formed my research, I relied on citation. In the submitted excerpt, I use feminist critiques of objectivity to recognise that the tool for knowledge production is the researcher herself. All knowledge is produced from a specific location that has to be discussed – I as a *tool* has to be accounted for in order for it to be a *good* tool.

---

6 An important point to note here is that I am in no way arguing for a hierarchy between 'native' and otherwise knowledge – the idea that everyone needs to only study their 'home' is not only impossible, but would also remove the majority of excellent works in general, and on the Balkans in particular. But what I am arguing for here is that whatever our location might be, we should ask ourselves why we ask the questions that we do, and how the tensions that inevitably arise from messy entanglements between us and our objects of study influence our research – not to lead us to paralysis, but to probe ever more.

At the same time, I claim the thesis is ultimately not about me. I am not the *object* of research. I am not Dauphinée's (2013) troubled researcher in centre stage, but I am the *voice* that needs situating. And I do this 'situating' in the thesis: I explain the curious situation of having been born in the same country as my interlocutors, but having now a Croatian passport that crosses borders far more easily than a Serbian one. I highlight the strange feeling of 'going home' to do fieldwork in a place where I understand the language and laugh at the jokes but a place that is not 'my country.' I do not write 'they' or 'the author' when describing my thoughts or actions. But I also do not start my chapters with stories.

When I asked myself why I used citations rather than stories, I first explained it by wanting to protect myself from a particular vision of failure. With the viva looming over my writing at the time of those final edits, I began to imagine horrible scenarios: failure of methodological rigour, failure of 'legitimate' knowledge production, failure to conform to the requirements of a successful Ph.D. thesis. What if the examiners do not see the connection between my stories and my theory? What if they say it has nothing to do with intervention? Instead, I presented my arguments through other people's words: by citing anthropologists, ethnographers, and IR scholars working on issues of positionality.

The anxieties turned out to be unfounded: my examiners were satisfied by the final product and I could slowly start to return to my writing with curiosity instead of dread. Seeing my writing with new eyes, I started resenting my own explanation of the cuts: it was not just fear of examiners that made me edit my thesis so ruthlessly. Saying this would be unfair to my examiners (who were open to different forms of writing), and to a generation of scholars who fought to create spaces within IR where authorial positionality is not only permitted but welcomed as an important analytical tool. How then, to explain the difference in the texts?

A threat to scientific legitimacy or a lack of comfort?

There are several ways to make sense of my writing decisions. My pre-submission anxieties are recognisable because they refer to the all-too-familiar ideals of detached, impartial, and objective science. They are rooted in the rigid parameters of 'objectivity' that have been so well picked apart by feminist scholars (Harding 1992; Haraway 1988; Rose 1997). Writing ourselves into our texts necessarily challenges these powerful ideals and can indeed be a scary endeavour.

The academe, however, has come a long way in the last few decades.

Although it is far from being radically transformed, there are spaces today for storytelling, autoethnography, and narrative, both outside of and within IR.[7] Moreover, I have read and been inspired by this literature, I see it as something to aspire *to*. No, I do not think the fear of the 'I' delegitimising my text can explain my decisions.

Why, then, does the first sentence of the submitted text reproduced above read like 'the soul of [my] writing [was] eviscerated, [my] passions sucked into a sanitised vortex that squeezes the life out of the things [I] write about' (Doty 2004, 380)? What keeps me attached to such faceless writing if it is not the desire to conform to academic requirements?

In the rest of this chapter, I discuss less explored reasons for choosing not to write ourselves in: it is not *always* or not *just* because we are afraid of failing to meet academic standards of detachment and objectivity. I am convinced of the urgency of changing the ways we deal with authorial presence within IR, and of the analytical and pedagogical potential of narrative writing. Precisely because I recognise this potential, this chapter discusses the difficult roads early career scholars have to navigate when deciding to use ourselves in writing.

Intimacy, authority, and the neoliberal academy

Even though I confidently state in the thesis that the 'research is not about me', I now see that I was attached to the vignette precisely because 'myself' looms in the background not only as a voice, but as an interesting *object*: one who comes home, one who is strangely bored by genocide, one who is surprised by Svjetlana's talk and wants you to follow the same lines of thinking, to ask the same questions, and to believe the answers provided. It is me as an *object* of interest, not just a tool or a voice, that is expected to lead you there.

The 'good' writing I strive for in these vignettes does two things. First, I work on developing a particular style that recuperates some of the 'sounds, rhythms, texture, and energy' (Doty 2004, 382) lost in academic texts. I want to guide the reader through my thinking without them tripping over long words, technical jargon, and new concepts. Second, I hope to provide a context of human experience: I create a backdrop in which I am meeting a friend for a night out in Belgrade, using social media, thinking about the people I see sitting around me. This experience is meant to 'recover human

[7] This literature is by now too large to review in a short space. Good introductions and overviews are provided by Carolina Moulin (2016), and Elizabeth Dauphinée and Paulo Ravecca (2018).

connection' (Doty 2010) and prop up my theorising that is to come – I want to show the reader that life happens outside statebuilding interventions.

In their discussion of narrative IR, Elizabeth Dauphinée and Naeem Inayatullah (2016, 2) admit they do not really know *how* these narratives do the things they do, but they know that they depend on *intimacy*. Intimacy is supposed to help in 'constructing and embodying an extensive architecture of understanding' (Inayatullah and Dauphinée 2016, 2). And intimacy is what I wanted to make visible in the cut out vignette: intimacy between me and my friends who took me to events and explained things, intimacy between me and the issues discussed that comes from my particular background, and, finally, the intimacy between me and the reader who is invited into this world where I can admit to being bored by genocide.

In making sense of my own writing decisions – execution of style, voice, and presence – I see that *intimacy* in the text scares me. But, I do not (just) fear intimacy *delegitimising* my text. With the pressure of the *viva voce* removed, I recognise I am also afraid of how attractive it might be – or, more precisely, how *I* might use this intimacy for obtaining very particular versions of success. Here, I have two particular versions of success in mind: 'I' as a successful commodity in neoliberal academia, and 'I' successfully occupying a position of an unquestionable authority.

In this call for intimacy that would rebuild human connections and give shape to abstract concepts, I am invited to present my friends, my family, and, ultimately, my home, as a thing to be consumed by academia. My theorising comes from a particular biography and learning how this biography shapes my reading and writing is an unfolding process – an exciting route of discovery and re-evaluation. But to really put myself in the text would require much more than changing the rhythm and texture of my writing. It would require explaining why someone with a Croatian passport would research Serbia. I would have to dwell on how Yugoslavia can mean something to people who have no real memories of it. I would need to discuss the complicated ideas of home, migration, and politics – ideas that require me to bring in not just myself, but those I am closest to. I would have to tell the story of myself and everyone who makes me.

Ideally, presenting this biography as a part (or a start) of our theorising has a 'purpose' – it helps fight against the 'single story', or helps IR to grasp complexity, fracture, and relationality (Ravecca and Dauphinée 2018). Someone might learn something about political transformations and the fate of people's hopes and dreams within them. The cynic in me, however, disagrees – in the best-case scenario, we shape students in our classrooms.

And, if our writing is just right, we *might* shape 'the discipline'.

In the context of early career precarity, this intimacy is also necessarily strategic. It becomes a part of publication plans and CVs – I might be rewarded with interest, citations, or even jobs. It is true that there is still a long way to go to turn the cracks in 'fortress academia' into liveable spaces where different types of writing are not considered a liability (Ravecca and Dauphinée 2018, 3). And there are no promises that the publications and jobs would really follow an attempt at narrative writing. But as a student of politics, I am painfully aware of the commodifying impulse that underlines these attempts, regardless of their outcomes. Whether my stories remain unread and forgotten, or become reviewed and cited, they have been turned into something that is evaluated by the metrics and standards of neoliberal academia which thrives on exclusion. *I* have turned my stories – and everyone who lives in them – into something to be consumed. Is it not treason to give our 'desires and wounds' (Inayatullah 2011, 6) to a ruthless system?

More than just citations and jobs, I would also be claiming a wholly different legitimacy: one that valorises personal narrative as a new authority. Despite my good intentions and the 'innocence in opening [myself]' (Doty 2004, 387), narratives are still powerful and seductive ways of ordering the world in a particular way (Wibben 2011, 2).

This move of 'ordering' according to knowledge accessible only to the narrator has been widely discussed in reflections on storytelling and narrative scholarship (Dauphinée 2016; Disch 2003; Ferguson 1991). As Megan Daigle (2016, 39) put it, the question is: 'By opening the door to the "I", do we lay out the welcome mat for any "authentic" experience – without further discussion?' Contrary to those who see narrative writing as closing off critique, the answer is no – narratives, whether ours or our interlocutors', do not imply a 'resurrection of a king unassailable standpoint epistemology' (Ravecca and Dauphinée 2018, 2, 4). On the contrary, narrative should be coming to terms with partiality and fracture, and a move to open, rather than close, oneself to critique (Dauphinée 2016, 51, 52). Its political potential lies precisely in *disruption* of the 'imaginary wholeness' of linear narratives (Edkins 2013, 288).

The democratising potential of narrative scholarship is thus said not to come from texts and authors less prone to 'authoritarian or reifying tendencies', but because they open up a space for (and depend on) 'the reader's active intervention' (Ravecca and Dauphinée 2018, 11). The 'readers' intervention', however, does not happen in a vacuum. The readers of my thesis have a well-defined image of the Balkans, they have detailed academic maps of characters and processes in which my interlocutors fit. I hoped to write against these representations and maps. And in fighting *against* these representations, positions of 'unassailable authority' become attractive.

Scholarship on the Balkans is overcrowded with representations, abstractions, and surprisingly resilient stereotypes. I admit to being tempted by a scenario in which my interpretation, even if its point is ambiguity, would *have to* be accepted, would over-write thousands of books on ethnic hatreds, would prevent anyone from unproblematically using the term 'civil society', and would banish forever the term 'transition' from graduate seminars. The question here is not about the *nature* of narrative writing, but whether *I* will be strong enough to resist ending tiring conversations with a simple and accessible claim to authenticity.

What do you think, then, would be a more insulting failure – that these stories that use love as a strategy invoke contracts and fame, or that they remain unread and ignored?

End note

Authorial presence is not an easy solution, but a gateway into complex negotiations. These negotiations need to address more than the 'self-indulgence' of the author and the 'orientalist exoticism' of an audience that might be forcing 'foreigners to write about themselves' (Inayatullah 2011, 7–8, 2–3). They also need to address the attractiveness and political use of narrative as authority, and its market value in the neoliberal academy. Politics of voice do not end with the inclusion of ourselves, on the contrary![8]

I do not advocate more 'fictive distancing' (Inayatullah 2011, 5) or a return to 'fortress writing' (Ravecca and Dauphinée 2018, 3). It would be counter-productive to offer any 'solution' at all. But in presenting a story of my own negotiation of these politics, I call for a more careful consideration of 'I' as an object, voice, and tool – not only what it can bring to our scholarship (as many have done), but how it might entwine with new and existing hierarchies. In presenting our souls to create a different kind of IR, we can easily forget that our 'souls' might be the last thing that early career researchers can protect from the market of academia. Consuming them uncritically might mean creating another unquestionable source of authority. Thus, instead of offering a 'how to', perhaps this lack of a conclusion is my way of becoming comfortable with some of the defining features of narrative writing: existing in/with ambiguity and abandoning the responsibility for closure (Inayatullah

[8] Himadeep Muppidi makes a similar point: https://thedisorderofthings.com/2013/03/23/reflections-on-narrative-voice/.

2013). Instead of referencing works of anthropology and IR or offering a defence, like I do in the thesis, I hope we talk about the different failures that emerge from our writing *and* the contexts in which we write.

* *The author would like to thank the workshop discussants and participants for their thoughts on the initial idea for this chapter. This chapter also benefitted from Jakub Záhora's reading and comments. All remaining mistakes are mine.*

References

Basberg Neumann, Cecilie, and Iver B. Neumann. 2015. 'Uses of the Self: Two Ways of Thinking about Scholarly Situatedness and Method'. *Millennium* 43 (3): 798–819.

Clifford, James. 1988. *The Predicament of Culture: Twentieth-Century Ethnography, Literature, and Art*. Cambridge, Mass: Harvard University Press.

Daigle, Megan. 2016. 'Writing the Lives of Others: Storytelling and International Politics'. *Millennium* 45 (1): 25–42.

Dauphinée, Elizabeth. 2013. *The Politics of Exile*. Interventions. Abingdon, Oxon: Routledge.

———. 2016. 'Narrative Engagement and the Creative Practices of International Relations'. In *Reflexivity and International Relations: Positionality, Critique and Practice*, edited by Jack L. Amoureux and Brent J. Steele, 44–60. New York: Routledge.

Disch, Lisa. 2003. 'Impartiality, Storytelling, and the Seductions of Narrative: An Essay at an Impasse'. *Alternatives: Global, Local, Political* 28 (2): 253–66.

Doty, Roxanne Lynn. 2004. 'Maladies of Our Souls: Identity and Voice in the Writing of Academic International Relations'. *Cambridge Review of International Affairs* 17 (2): 377–92.

Edkins, Jenny. 2013. 'Novel Writing in International Relations: Openings for a Creative Practice'. Security Dialogue 44 (4): 281–97.

Ferguson, Kathy E. 1991. 'Interpretation and Genealogy in Feminism'. *Signs* 16 (2): 322–39.

Geertz, Clifford. 1973. *The Interpretation of Cultures: Selected Essays*. New York: Basic Books.

Gordon, Edmund T. 1997. 'Anthropology and Liberation'. In *Decolonizing Anthropology: Moving Further toward an Anthropology of Liberation*, edited by Faye Venetia Harrison, 149–67. Arlington, Va.

Gupta, Akhil, and James Ferguson. 1997a. 'Culture, Power, Place: Ethnography at the End of an Era'. In *Culture, Power, Place: Explorations in Critical Anthropology*, edited by Akhil Gupta and James Ferguson. Durham, N.C: Duke University Press.

———. 1997b. 'Discipline and Practice: "The Field" as Site, Method, and Location in Anthropology'. In *Anthropological Locations: Boundaries and Grounds of a Field Science*, edited by Akhil Gupta and James Ferguson. Berkeley: Univ. of Calif. Press.

Haraway, Donna. 1988. 'Situated Knowledges: The Science Question in Feminism and the Privilege of Partial Perspective'. *Feminist Studies* 14 (3): 575–99.

Harding, Sandra. 1992. 'Rethinking Standpoint Epistemology: What Is "Strong Objectivity?"' *The Centennial Review* 36 (3): 437–70.

Inayatullah, Naeem. 2011. 'Falling and Flying: An Introduction'. In Autobiographical International Relations: I, IR, edited by Naeem Inayatullah, 1–12. London: Routledge.

———. 2013. 'Distance and Intimacy: Forms of Writing and Worlding'. In *Claiming the International*, edited by Arlene B. Tickner and David L. Blaney. Milton Park, Abingdon, Oxon: Routledge.

Inayatullah, Naeem, and Elizabeth Dauphinée. 2016. 'Permitted Urgency: A Prologue'. In *Narrative Global Politics: Theory, History and the Personal in International Relations*, edited by Elizabeth Dauphinée and Naeem Inayatullah. London: Routledge.

Lie, Jon Harald Sande. 2013. 'Challenging Anthropology: Anthropological Reflections on the Ethnographic Turn in International Relations'. *Millennium* 41 (2): 201–20.

Mani, Lata. 1990. 'Multiple Mediations: Feminist Scholarship in the Age of Multinational Reception'. *Feminist Review* 35: 24–41.

Moulin, Carolina. 2016. 'Narrative'. In *Critical Imaginations in International Relations*, edited by Aoileann Ní Mhurchú and Reiko Shindo, 136–53. London: Routledge.

Ravecca, Paulo, and Elizabeth Dauphinée. 2018. 'Narrative and the Possibilities for Scholarship'. *International Political Sociology* 12 (2): 125–38.

Rich, Adrienne. 1986. 'Notes Towards a Politics of Location'. In *Blood, Bread, and Poetry: Selected Prose, 1979–1985*, 210–31. New York: Norton.

Rose, Gillian. 1997. 'Situating Knowledges: Positionality, Reflexivities and Other Tactics'. *Progress in Human Geography* 21 (3): 305–20.

Van Maanen, John. 2011. *Tales of the Field: On Writing Ethnography.* Chicago: University of Chicago Press.

Vrasti, W. 2008. 'The Strange Case of Ethnography and International Relations'. *Millennium* 37 (2): 279–301.

Wibben, Annick T. R. 2011. *Feminist Security Studies: A Narrative Approach*. London: Routledge.

Concluding Reflections

13

Building on Ruins or Patching up the Possible? Reinscribing Fieldwork Failure in IR as a Productive Rupture

BERIT BLIESEMANN DE GUEVARA AND XYMENA KUROWSKA

Introduction

Fieldwork research hardly ever goes as planned. Struggles around issues such as getting access to research participants or what participants are willing to share are often frustrating to the extent that they generate feelings of failure in the researcher. Sometimes fieldwork realities make it impossible to carry out projects as originally proposed. We have certainly been there; indeed, most fieldworkers have. Fieldwork failure in IR should thus be business as usual or even an opportunity for research breakthroughs. It should be somewhat akin to Karl Popper's idea of falsification (Popper 1959), narrative scholarship's notion of surprise as opportunity to enrich analysis (Ravecca and Dauphinée 2018), or the interpretivist idea of blunders as a way to reconstruct social meaning in the field (Shehata 2006). But we rarely experience failure as part of experimentation or productive opening. If something goes wrong, it is not an occasion to learn but a reason to despair.

This is a logical response to the fact that failure, including fieldwork failure, is an academic taboo. The discourse of the neoliberal university presupposes control and glorifies success and its quantification. Openly admitting or embracing fieldwork failure in this context would mark a breach – with tangible reputational damage for the researcher and her university. Fieldworkers are caught in a web of structural, epistemological, and situational contradictions. Though fieldwork was an unorthodox research strategy for IR scholars as the discipline came of age, it is now in high demand. Despite this, training remains scarce. An exciting new opportunity to widen the methodological, ethical and analytical horizons of IR, fieldwork is often circumscribed by the discipline's overriding empiricism (Vrasti 2008). Its instrumental approach, based on the logic of data extraction, blends well with neoliberal demands for entrepreneurship. Fieldwork supplies evidence as the researcher leaves the proverbial ivory tower to get her hands dirty in an effort to generate not only data but also impact. It has thus become a staple of grant applications. The double quality of a greater engagement with those who 'do' everyday international politics, on the one hand, and the danger of having this engagement hijacked by neoliberal logics, on the other, is also visible in the process of institutional ethics clearances which aim to ensure both the safe treatment of human subjects and the researcher, but which also aim to safeguard the university from liability. This process is based on a 'duty of care' principle which is often inadequate when research takes place in violent and/or illiberal contexts (Russo and Strazzari, 2020).

In short, both a political economy and an academic cottage industry have consolidated around fieldwork. Although a thoughtful, immersive, hands-on literature on fieldwork as practice of knowledge production is growing (Glasius et al. 2018, De Goede, Bosma, and Pallister-Wilkins 2019, Daigle 2017, Bliesemann de Guevara and Bøås, 2020), there is less sustained reflection on what the demand for fieldwork means for academic subjectivity. Looking fieldwork failure in the eye is a productive locus from which to start such reflection. After all, failure is not a correctable obstacle but shows certain ideology at work (Straehler-Pohl and Pais 2014). It marks a moment of dislocation and a sense of displacement, which exposes a set of relations that are usually hidden or subdued. When recognised as such, it confronts us with the darker corners of IR life.

In this concluding chapter, we contextualise fieldwork failure sociologically and reinscribe its meaning. This is not a consolatory tale, and we do not excuse the researcher from the responsibility to exercise craft and due care in their fieldwork. Examining the status of failure helps, however, to integrate the politics of the discipline with the politics of the field beyond merely blaming the researcher. We suggest in particular replacing failure with 'productive rupture' as a useful overall denominator and consider specific ways of reinscribing failure in different dimensions of fieldwork. We do not aim to haughtily transform failure into success. We rather want to put failure in its place by understanding how it structures IR life. In order to do so, we first provide a vignette from one of our projects, and second bring to bear four socio-political facets of fieldwork failure in IR: structural and epistemological conditions, contingency, the anxiety generated to a large extent by the overlap of the first two, and the privilege to fail which manifests stratification within the academy.

It is from the position of relative academic privilege that we examine our record of failure as part of the scholarly endeavour of fostering debate, without, we hope, precipitating a downfall. The background to this reflection is our own recent bumpy field research on Polish border guards narrated elsewhere (Kurowska 2019b) and former guerrilla fighters in Colombia described in the vignette below. Surely, we first 'failed' in fieldwork during the doctoral projects for which, back in the day, nobody trained us. Both of us look back on them in some horror for having done everything wrong from our present-day perspective and despite the validation of the doctoral thesis (Kurowska 2019a). There was no space in IR to talk about such trajectories then and it would have been reckless for early career scholars to even try. Given these hierarchies, the courageous probing of failure that the mostly early career contributors to this volume undertake stands out. They go beyond wearisome prescription or declaration towards embodied reflection. They have been there and they take the reader with them through their engaging writing. They also confront the taboo that would rather have them, as used to be and often remains the case, first, pretend that the failure did not happen; second, re-design in private projects gone awry; and third, bounce back into the posture of control. They instead take on the politics of fieldwork failure in the life of IR.

Vignette: How things go wrong (and then deliver)

Among the reasons for fieldwork failures, changing circumstances and participants' agency – exemplified by the following observations from research on the subjectivities of former Colombian armed actors in the process of reincorporating into civilian society – are very common. Field research is dependent upon contingent contextual circumstances beyond the researcher's control. A change in circumstances may create uncertainty and require adaptations. In the case of the Colombia project, fieldwork involved negotiating research access to political prisoners of the guerrilla group ELN (*Ejército de Liberación Nacional*), which was holding peace talks with the Colombian government. Through a tedious bureaucratic process, we successfully obtained the prison management's written permission to work with ELN prisoners, most of whom were keen to participate in our biographical conversations and needlework. When our fieldwork commenced some weeks later, however, the political context had changed considerably: The ELN had claimed responsibility for a car bomb attack on a police academy in Bogotá, the peace talks had broken down, and public discourse had made a marked shift towards securitisation. We were denied access to the ELN prisoners by the very officials who had granted it before, claiming that there were 'no political prisoners' but only 'terrorists'; that only social interventions, not research, were of interest to the prison; that such interventions should target all prisoners, not just a specific section; and that as a prison the peace and reconciliation process was not of their concern. Despite all players remaining the same, the change in socio-political circumstances frustrated our research access and denied the rank-and-file political prisoners the hope of finally being listened to. After our initial project presentation, we were asked by them whether we were really committed to working with them. We affirmed truthfully but also emphasised that access ultimately depended on the prison. Why did this still feel like a major failure, like we were letting these men down?

Another common cause for failed fieldwork is participants being reluctant or refusing to partake. In view of the closed ELN route, we intensified our access negotiations with members of the demobilised guerrilla group FARC (*Fuerzas Armadas Revolucionarias de Colombia*), which had signed a peace agreement with the government in 2016 that included government support for former fighters to settle in Territorial Spaces of Capacitation and Reincorporation (ETCR) – obvious locations for our fieldwork. We met with ETCR leaders, FARC political party representatives, FARC women's organisations, and others, but the result was always the same. Despite general interest, these gatekeepers emphasised more pressing needs. At the time of our conversations, 128 FARC leaders had been killed by paramilitary or criminal forces (the number has risen further since); many ETCRs had already dissolved due to the state's failures in implementing the peace agreement; still-existing ETCRs were struggling to become economically sustainable. The gatekeepers were thus keen to invite 'productive projects' to create socio-economic opportunities. Frustration caught hold of our team: Why couldn't our interlocutors see that our project aimed at unstitching the securitised, binary public discourse that contributed to their situation in the first place? How could we convince them to prioritise their communities' wellbeing over gatekeeper power games? Why had funding bodies in Colombia and the UK issued a research call on reconciliation, when there was little to reconcile, and why had we given in to the neoliberal pressures to apply for such funding? After six months of fruitless conversations, the project was on the verge of collapse due to a lack of participants able (ELN) or willing (FARC) to partake. Against the background of a Colombian funding body which makes the individual Principal Investigator, not their institution, financially liable for 'failed' projects, aborting the research became a real option to minimise damage.

A month later, we finally started fieldwork in a New Reincorporation Point (NPR), a community of former fighters and associated civilians who had pooled money to buy land on which they were building a village, largely without public support. What ultimately changed the tide was persistent discussion of our failures with colleagues, and our continued search for openings in the protracted social and security situation. We were introduced to this NPR community by an academic colleague upon telling him about our problems. Likewise, upon hearing about our failures, the Peasant Association of Antioquia, with whom some team members had long-standing working relationships, contributed a 'practical' element in form of a voluntary agrarian advisor to our project. This reciprocity and kindness, emanating from trusting relationships and talk (see Danielle House's chapter in this volume), helped us gain access to an NPR community that embraces and owns our project. Were the initial fieldwork failures – as openings in our strategies and imaginations – necessary to end up here? And should we brace for further frustrations, further failures? After all, we know now that it would only take one act of violence, confrontation with state authorities, or unethical colleague to lose the hard-won access and push the project back to the brink of failure.

What failure manifests

Why is it so difficult to incorporate fieldwork failure into what we do (write, speak, teach) as IR researchers? Why do we even continue to use the misleading term 'failure', if ruptures may deliver more generative engagement with the social reality we study (see Lydia C. Cole's chapter in this volume)? To address these questions, it is useful to come back to the overarching themes of this chapter and volume: What is fieldwork failure and why is it taboo? What goes missing when we fail to examine failure? Can failure be revelatory, despite the heavy emotional labour and professional hazards that come with it, or must we rather resist the ideology of success that makes it necessary to turn failure into a productive moment? Does failure enable traversing the strictures of the academic frame, learning rather than only testing something? The following socio-political facets of fieldwork failure in IR contextualise our own and others' 'failure' to speak to these themes.

Structural conditions and epistemological script. Fieldwork failure exposes a particular academic subject position which is shaped by the discourse of mastery marked by 'the will to know' (Foucault 2013), accomplishment, and status (Bourdieu 1990). As researchers, we are socialised into, and expected to represent, such discourses, both in and out of the armchair (see Jan Daniel's and Renata Summa's chapters in this volume). We are supposed to know before we get a grip of what there is to know, and control the process of bringing such knowledge to bear. This is partly a legacy of IR adopting the natural science convention of 'writing from' a successful experiment and erasing the messy process of experimentation behind the scenes (Latour and

Woolgar 1986). Such façades strengthen the perception of an easy fit between data and research, 'leaving little trace of the agony and uncertainty of construction' (King, Keohane, and Verba 1994, 13). We are, as a result, caught in the 'organised hypocrisy' of the research design. To have the research design approved or stand a chance of 'grant capture', we need to demonstrate that we have already mastered the field and can therefore offer 'value for money'. This structurally-induced strategy upends the idea that a mapped-out field is an outcome, rather than a preparatory step of a research project (Amit 2000).

From within such a subject position, fieldwork failure feels like personal responsibility, and the researcher suffers the neoliberal pain and shame of perpetual inadequacy (see Johannes Gunesch and Amina Nolte's chapter in this volume). While this is a standard way of regulating conduct in the Western academy, the bureaucratic manifestations of quality assurance can take particularly punitive forms when, as in the Colombia project, the Principle Investigator is financially liable if the project does not meet its objectives as stated in advance and meticulously laid out in a detailed work plan. Some of the creative 'solutions' the Colombia project team developed in view of looming project failure – such as 'following' former inhabitants of dissolved ETCRs to their new homes, often located in Medellín's shantytowns – were turned down by the Colombian funder on the grounds that they did not meet the geographical parameters of the original call, even though it was written in a politically different situation from today's. The best strategy may in this context be to adopt contextually appropriate micro-tactics (see also Steele, Gould, and Kessler 2019) and strengthen solidarity among individual researchers on the team. The Colombia project team, for instance, secured additional, smaller but more flexible funding and made new connections which allowed them to complement their work in the NPR with fieldwork in other locations, and with other groups not covered by the rigid grant.

Contingency, or circumstances. Technically perfect research designs attract funding but crumble in the field (see Holger Niemann's chapter in this volume). Access gets routinely denied despite purposefully cultivating relevant relationships and approaching gatekeepers with finesse. Even where it has been granted at one point, political dynamics – the very reason why the project is conducted in the first place – may slam the window of opportunity shut at any time. Defiant interlocutors do not want to give us what we envisage they owe us. In fact, even if they formally abide, 'forced' rather than negotiated access rarely generates rich data (see Desirée Poet's chapter in this volume). By exercising their right to information as specified in national and international legal provisions, the researcher risks antagonising relevant participants and may be forced to settle for redacted documentation. This is a blind ally: We end up with partial information without a chance to make sense of it within a conversation.

Secrecy in the case of security agencies is particularly frustrating as such establishments can afford to deny access without justification by invoking national security. Being denied access in such situations feels like an ethical slight too, as such institutions should, after all, be accountable. Their blatant rebuff and refusal to abide is only the most visible and ritualistically decried manifestation of the researcher's lack of control. Interlocutors may also refuse to engage because they (rightfully) decide that the researcher's concerns, which may be interesting in general, are not their most pressing problem.

If the structural conditions of neoliberal IR force failure upon the researcher, contingencies in the field ought to feel like a failure precisely because we enter the reality of the other. We encounter difference as disoriented strangers (see Ewa Maczynska's chapter in this volume). Approached in this way, fieldwork failure obviously hurts, too; but it hurts differently. Mediated by the acknowledgement that the other is not obliged to talk to us, failure is a rite of passage in the process of making sense of a new place. In the spirit of the ethnomethodological tradition, the researcher may even seek to commit what she senses are social blunders. While reactions to them are uncontrollable, much can be learned from 'purposeful' breaches of local social rules (Garfinkel 1967).

Improvisations require practice, however, methodological as much as emotional, and a disposition to bear such situations. This attitude produces its own vulnerability, but it differs from the neoliberal sense of inadequacy. In embracing contingency, the researcher consciously takes on a role of a supplicant and confronts her own epistemic hubris. The failure to get a joke by an interlocutor is an opening, rather than a closure (Rose 1997). A useful way of reinscribing this facet of fieldwork failure is to think of it as exposure (Schwartz-Shea and Yanow 2012, 85) to a wide variety of meanings by the interlocutors, which may be both contradictory to each other as well as to the researcher's interpretive frame. This, again, is not a threat, but an opportunity for 'thickening' our interpretations.

Anxiety. The intense emotional charge to fieldwork failure transpires at the juncture of the structural pressures and pressures of contingency which bring in the neoliberal shame and angst of encountering the other, on the one hand, and the researcher's idiosyncratic disposition and the affect of the moment, on the other. In fieldwork, as the ethnographic tradition teaches us, the researcher is her own research instrument and there is no shelter from the state of anxiety. The emotional charge cannot be defused because it is inherent to fieldwork, rather than a side effect to be mediated away (see Jakub Záhora's chapter in this volume). We may seek to protect ourselves from it by 'omission, soft-pedalling, non-exploitation, misunderstanding, ambiguous description, over-exploitation or rearrangement of certain parts of [our] material' (Devereux 1967, 44). This obviously poses analytical dilemmas. Much of interpretive work materialises through emotions, but emotions enable and undercut our interpretive powers at the same time. They make us juggle the necessity to seek and make sense of discomfort and the responsibility to protect ourselves. Some constellations of anxiety will prove unbearable: We may have to leave the field and come back later, or not at all.

The emotional strain finds its own form and risk of failure in (post-)conflict environments. Shesterinina's (2019) reflection on her avoidance of some former combatants, whom she feared, involved an imposition of moral schemes and ultimately a flattening of her understanding of participation in violence (see Emma Mc Cluskey's chapter in this volume). Some researchers in the Colombia project team had a knee jerk reaction against talking with former right-wing paramilitaries known for their grave human rights violations in the 1990s and early 2000s. What would it mean, however, to work with such interlocutors despite our fear or moral aversion (see Sezer İdil Göğüş's chapter in this volume)? We may, as narrative scholarship suggests (Ravecca and Dauphinée 2018), surprise ourselves with how multi-layered the stories of such others are, and how our own expanded range of reactions enriches the analytical narratives we construct. Generative as it promises to be, reinscribing anxiety as capacity for surprise in fieldwork failure remains a challenge. The simultaneous multiplicity, tangibility and indiscernibility of emotions renders vulnerable all the parties to the fieldwork conversation and alludes prescription.

Privilege. Few can afford to fail. Even fewer are in a position to admit and explore failure as an academic project. Junior scholars who venture into those terrains normally enjoy the mentorship of established critical scholars and the support of renowned critical research programmes, which can turn failure into the next cutting-edge debate (see Katarina Kušić's chapter in this volume). We did not experience that advantage at the beginning of our careers and might not have been able to interrogate failure had we not met each other and the community in which such discussions are possible. How productive fieldwork failure will be depends therefore in part on academic hierarchies, lineage, and disciplinary geographies of eligible innovators. Scholars not based in renowned academic sites, regardless of the stage of their career and quality of their research, may tend to have their 'failures' considered as a lack of professionalism. Structurally, this reflects the fact that failure is a privilege. The academic 'class', somewhat overlapping but not identical to the class structure outside of academia, inscribes itself in the politics of fieldwork failure.

Reproduction of privilege remains entrenched in the politics of academia, but can be at least partly reinscribed through insistence on incorporating positionality into our research claims. Reflexivity through positionality considers more holistically the researcher's role in the construction of the research problem, including her social and ideological standing, and thus exposes how knowledge is marked by its origins (Lynch 2008). It finds certain limits, however, in the cognitive pressure to produce outputs in short timeframes and with tangible results, such as peer-reviewed articles in highly-ranked journals. This structure in turn fuels the failure taboo.

Conclusion: Reinscribing failure

Failure is produced structurally but experienced personally, and is always hard to take. It thrives on and feeds the imposter syndrome, which in turn blossoms on the glorification of success in the academy. All academics, by virtue of being academics, are haunted by it. We struggle with the effects of such discourses, as we also argue that the real failure is to not problematise the academic frame of mastery and the failure taboo. We hope that the four facets of fieldwork failure that we identify, together with the associated strategies of reinscription – micro-tactics, exposure, the capacity for surprise, and reflexivity through positionality – help rupture failure and put it in its place by opening spaces for communication. This is by no means an exhaustive or authoritative categorisation, but one which results from our own experiences.

The two metaphors in the title, 'building on ruins' and 'patching up the possible' reflect our respective field research strategies as they have developed since our doctoral projects. They conjure up different, if related, imageries of failure and its implications, and relate to a larger question of resilience. Resilience has had a bad press in some critical corners as yet another manifestation of self-regulation on the part of the neoliberal subject who always bounces back, gets back to the grind, and makes the most of it for the system. We have been resilient, even tenacious, as we have learned to rupture failure and thus subvert the limitations of the neoliberal subject. This is an idiosyncratic process where much strength and inspiration comes from talking and working with others, both co-researchers and interlocutors. Reciprocity of good relationships is what has often carried us through, emotionally and as a way of handling interpretative and logistical impasses. We find that conversation helps rupture failure, but this will not be a solution, and not a strategy for everyone. However, as an expression of camaraderie and solidarity this chapter and this volume will hopefully encourage others to break the failure taboo, too.

**Bliesemann de Guevara's contribution draws on the international* *collaborative research project "(Des)tejiendo miradas sobre los sujetos en proceso de reconciliación en Colombia/(Un)Stitching the subjects of Colombia's reconciliation process", supported by Colciencias (project reference FP44842-282-2018) and the British Newton Fund (project reference AH/R01373X/1), and hosted by the University of Antioquia, Colombia, and Aberystwyth University, UK (2018–2020). Xymena Kurowska's work on this chapter was funded through European Commission MSCA Individual Fellowship RefBORDER grant no. 749314.*

References

Amit, Vered, ed. 2000. *Constructing the field: ethnographic fieldwork in the contemporary world*. London and New York: Routledge.

Bliesemann de Guevara, Berit, and Morten Bøås, eds. 2020. *Doing Fieldwork in Areas of International Intervention: A Guide to Research in Violent and Closed Contexts*. Bristol: Bristol University Press.

Bourdieu, Pierre. 1990. *Homo academicus*. Cambridge: Polity Press.

Daigle, Megan. 2017. 'Learning from the Field'. In *Routledge Handbook of International Political Sociology*, edited by Xavier Guillaume and Pinar Bilgin, 281–289. London: Routledge.

De Goede, Marieke, Esmé Bosma, and Polly Pallister-Wilkins, eds. 2019. *Secrecy and Methods in Security Research: A Guide to Qualitative Fieldwork*. London: Routledge.

Devereux, George. 1967. *From anxiety to method in the behavioral sciences*. Den Haag and Paris: Mouton & Co.

Foucault, Michel. 2013. *Lectures on the will to know*. Basingstoke: Palgrave Macmillan.

Garfinkel, Harold. 1967. *Studies in ethnomethodology*. London and Englewood Cliffs,N.J.: Prentice-Hall.

Glasius, Maries, Meta de Lange, Jos Bartman, Emanuela Dalmasso, Aofei Lv, Adele Del Sordi, Marcus Michaelsen, and Kris Ruijgrok. 2018. *Research, Ethics and Risk in the Authoritarian Field*. Basingstoke: Palgrave.

King, Gary, Robert O. Keohane, and Sidney Verba. 1994. *Designing social inquiry: scientific inference in qualitative research*. Princeton, N.J.: Princeton University Press.

Kurowska, Xymena. 2019a. 'When home is part of the field: experiencing uncanniness of home in field conversations'. In *Tactical Constructivism: Expressing Method in International Relations*, edited by Brent Steele, Harry Gould and Oliver Kessler. London and New York: Routledge.

———. 2019b. 'When one door closes, another one opens?: The ways and byways of denied access, or a Central European liberal in fieldwork failure'. *Journal of Narrative Politics* 5 (2): 71–85.

Latour, Bruno, and Steve Woolgar. 1986. *Laboratory life: the construction of scientific facts*. Princeton, N.J.: Princeton University Press.

Lynch, Cecelia. 2008. 'Reflexivity in Research on Civil Society: Constructivist Perspectives'. *International Studies Review* 10 (4): 708–721.

Popper, Karl R. 1959. *The logic of scientific discovery*. London: Hutchinson.

Ravecca, Paulo, and Elizabeth Dauphinée. 2018. 'Narrative and the Possibilities for Scholarship'. *International Political Sociology* 12 (2): 125–138.

Rose, Gillian. 1997. 'Situating Knowledges: Positionality, Refexivities and other Tactics'. *Progress in Human Geography* 21 (3): 305–320.

Russo, Alessandra, and Francesco Strazzari. 2020. 'The politics of safe research in violent and illiberal contexts'. In *Doing Fieldwork in Areas of International Intervention: A Guide to Research in Violent and Closed Contexts*, edited by Berit Bliesemann de Guevara and Morten Bøås. Bristol: Bristol University Press.

Schwartz-Shea, Peregrine, and Dvora Yanow. 2012. *Interpretive research design: concepts and processes*. New York and London: Routledge.

Shehata, Samer. 2006. 'Ethnography, Identity, and the Production of Knowledge'. In *Interpretation and Method: Empirical Research Methods and the Interpretive Turn*, edited by Dvora Yanow and Peregrine Schwartz-Shea, 244–63. Armonk, NY and London: M.E. Sharpe.

Shesterinina, Anastasia. 2019. 'Ethics, empathy, and fear in research on violent conflict." *Journal of Peace Research* 56 (2): 190–202.

Steele, Brent, Harry Gould, and Oliver Kessler, eds. 2019. *Tactical Constructivism: Expressing Method in International Relations*. London and New York: Routledge.

Straehler-Pohl, Hauke, and Alexandre Pais. 2014. 'Learning to fail and learning from failure – ideology at work in a mathematics classroom'. *Pedagogy, Culture & Society* 22 (1): 79–96.

Vrasti, Wanda. 2008. 'The Strange Case of Ethnography and International Relations'. *Millennium* 37 (2): 279–301.

Note on Indexing

E-IR's publications do not feature indexes. If you are reading this book in paperback and want to find a particular word or phrase you can do so by downloading a free PDF version of this book from the E-IR website.

View the e-book in any standard PDF reader such as Adobe Acrobat Reader (pc) or Preview (mac) and enter your search terms in the search box. You can then navigate through the search results and find what you are looking for. In practice, this method can prove much more targeted and effective than consulting an index.

If you are using apps (or devices) to read our e-books, you should also find word search functionality in those.

You can find all of our e-books at: http://www.e-ir.info/publications

www.ingramcontent.com/pod-product-compliance
Lightning Source LLC
Chambersburg PA
CBHW030246030426
42336CB00009B/283
* 9 7 8 1 9 1 0 8 1 4 5 3 6 *